BORN OF FIRE AND RAIN

M. L. HERRING

Born of Fire and Rain

JOURNEY INTO A PACIFIC COASTAL FOREST

Yale UNIVERSITY PRESS / NEW HAVEN AND LONDON

Published with assistance from the Mary Cady Tew Memorial Fund.

Yale University Press books may be purchased in quantity for educational, business, or promotional use. For information, please e-mail sales.press@yale.edu (U.S. office) or sales@yaleup.co.uk (U.K. office).

Designed by Mary Valencia.
Set in 12/16 Garamond Premier Pro type by Motto Publishing Services.
Printed in the United States of America.

Illustrations by the author.

Library of Congress Control Number: 2024933760
ISBN 978-0-300-27542-1 (hardcover : alk. paper)

A catalogue record for this book is available from the British Library.

This paper meets the requirements of ANSI/NISO Z39.48-1992 (Permanence of Paper).

10 9 8 7 6 5 4 3 2 1

This book is for you, Cole and Ellis, because the future is yours. My greatest hope is that someday you will walk among giant trees in a wild rainforest and see it for yourselves.

CONTENTS

PREFACE AND ACKNOWLEDGMENTS

If you are going to live on a rapidly changing planet, you'd be wise to learn how it works. And there's no better way to learn than to follow those who have lived here, through good times and not-so-good, for a very long time. The giant, old trees on a skinny stretch of land on the far West Coast of North America have a lot to say about living in a twitchy world.

No, this is not a story about California redwoods. This story is set a bit farther north, in a place that is often overlooked even by the people who live here, a place that will surprise you with its incredible beauty, rambunctious character, and tenacious will to live.

You are about to embark on an imagined expedition into America's temperate rainforest, at the tumultuous edge of a shifting continent in a precarious moment of time. You will peek behind the curtain of magnificent scenery into a forest of ancient trees, exploding mountains, disappearing owls, tsunamis, megafires, timber wars, salmon, epiphanies, ten million people, and a lifetime of drawing pictures. And that's just scratching the surface.

More than an episodic travelogue, this book takes you across scales from microscopic to planetary, underground, up in the air, and into places only your imagination can fit, to learn what it means to be a forest. At first you will be awed by the obvious majesty of these giant, moss-draped trees. You will witness the re-

---

Fundamental to the story of the Douglas-fir region is its location on the Pacific Ring of Fire.

gion's tumultuous human history and glimpse its possible future
in a world of climate change, species extinction, and shifting environmental policies. In the end, you will emerge with mud on your
boots and an expanded idea of what it means to be human in an
ecological world.

The illustrations in this book are from sketchbooks I have
kept most of my life, not as a photographic record but as a personal
response to the world as I have experienced it. The sketches offer an
alternate path into the rainforest, a slower path that allows you to
stop for a moment, just as you might pause along a trail to glimpse a
fleeting owl. At this particular moment, when the earth is shifting
all around, there's a lot to learn about life in a wild forest ecosystem.
I hope this imagined expedition will encourage you to explore the
natural wonders in your own nearby forest with new awareness.

I will be your guide on this journey, to share the knowledge
and insights I have gathered from years of following around some
very smart people. Because of these scientists, activists, and artists,
we still have a temperate rainforest to explore. Many thanks to Fred
Swanson, Tom Spies, Steve Sillett, William Robbins, Audrey Per-
kins, Brooke Penaluna, Kim Nelson, Kathleen Dean Moore, Penny
McDermott, Damon Lesmeister, Chandra LeGue, Barb Lachenbruch, Molly Keogh, Julia Jones, Sherri Johnson, Norm Johnson,
Mark Harmon, Josh Halofsky, Jerry Franklin, Gordon Grant, Paul
Englemeyer, Dave Buchanan, Mary Braun, Dan Bottom, Linda
Ashkenas, Matt Betts, Roger Worthington, and many, many more.

And thanks also to my careful, caring agent, Rachel Vogel,
who saw something in this story, even though she had never seen
this forest; and to the fine, thoughtful editors at Yale University
Press: Jean Thomson Black, Jeff Schier, and Elizabeth Sylvia, who
made it possible. Thank you all for sharing in the forest's story.

There's another long list of family and friends who offered me
ideas and encouragement. First among them is my husband, John
Keogh, who has supported this project since he first balanced on a
floating cedar log and built me a house in the forest.

BORN OF FIRE AND RAIN

Forest
of
Fire & Rain

Vancouver Is.

Salish Sea

Olympic
Mountains

Seattle

Tacoma

Mt.
Rainier

Mt. St.
Helens

Mt.
Adams

Pacific
Ocean

Portland

Mt. Hood

Cascade
Mountains

Eugene

Three
Sisters

Coast
Range

Mt.
Thielsen

Crater
Lake

Klamath
Mountains

I

YOU ARE HERE

A forest of these trees is a spectacle too much for one man to see.

—DAVID DOUGLAS, nineteenth-century British explorer

Here you are, standing halfway between the equator and the North Pole, squeezed between the Cascade volcanoes and the Pacific Ocean. Trees tower like skyscrapers above you, a carpet of moss softens your path, an evergreen scent wreaths the air. It's raining, of course. This is part of the Pacific temperate rainforest, the largest temperate rainforest in the world. Stretching from the giant cedars of southeast Alaska to the giant coast redwoods in northern California, this narrow ribbon of rainy forest has a distinctively benign climate for trees to grow very big and very old.

The Pacific temperate rainforest is completely unlike the drier, interior forests that spread across most of western North America. Giant conifers dominate this landscape, much like on Earth in the time of the dinosaurs. Conifers (cone-bearing trees) are a holdover from a time before flowering plants, before the rise of mammals, before the continents drifted into their current locations. The Pacific temperate rainforest is the only sizable region on Earth where conifers flourish as they did before the rise of flowering plants. When we walk among these trees, so huge and so ancient, it's easy to imagine a *Brontosaurus* browsing the mid-canopy among trees a thousand years old and ten feet in diameter.

You will explore a place of fiery origins, drenching rain, and a tenacious grip on life.

3

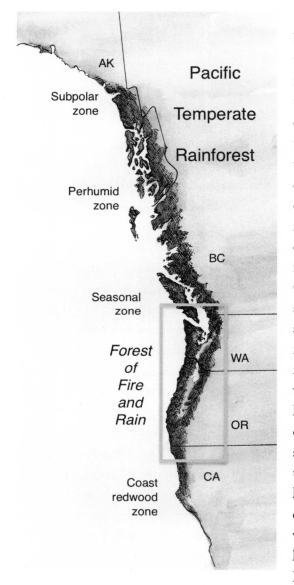

Characterized by mild temperatures and a lot of rainfall, the Pacific temperate rainforest stretches for two thousand miles up the North American coast, where it crosses several climate zones and ecoregions, all dominated by conifers. Our forest of rain and fire is the "seasonal zone" of the rainforest, dominated by Douglas-fir with western hemlock, western redcedar, and Sitka spruce. North of us, in British Columbia and southeast Alaska, is the wetter, cooler "perhumid zone" where Douglas-fir gives way to cedars, spruces, and hemlocks. To the south of us is the narrow "coast redwood zone" in northern California, where fog more

A "forest of fire and rain" describes the seasonal rainforest zone of the larger Pacific temperate rainforest.

than rain dampens the forest. Collectively, this Pacific temperate rainforest is packed with biomass—living and decaying trees,
shrubs, mosses, and soil—in volumes greater than any other ecosystem on the planet. And it stores massive amounts of carbon.[1]

More rare than *tropical* rainforests that cluster around the
equator, *temperate* rainforests cling to a few mid-latitude strips
along the coasts of the Northern and Southern hemispheres.[2] Some
of the world's tallest trees grow in temperate rainforests. Mountain ash (*Eucalyptus regnans*) in the highlands of Tasmania and
California's coast redwoods each can grow to over 325 feet, taller
than the Statue of Liberty and more than *four* times the height
of the tallest eastern white pine or sugar maple. These tall tree
species are critically endangered, whittled down by logging and
development.

There is another giant tree, iconic to this temperate rainforest: the coastal Douglas-fir. It too can exceed heights of 325 feet,
and it dominates the rainy forests of western Oregon and Washington, from the volcanic crest of the Cascade Mountains to the
edge of the Pacific Ocean. This is the sweet spot of the Pacific temperate rainforest, a place rainier than northern California and
more fire-prone than soggy coastal Alaska. This far west slope of
the Pacific Northwest has many names: the coastal Douglas-fir region, the range of the northern spotted owl, or simply the Westside. It shelters some of the world's biggest trees and most famous endangered species. Tucked between latitudes forty and fifty
north, this forest of fire and rain is never much more than one
hundred miles from the Pacific Ocean. It embraces traditional
lands of the Kalapuya, Salishan, Klickitat, and other Native people. And it is currently home to ten million people, most of whom
have no idea they live in a rainforest.

Sheltered between the world's largest ocean and the Pacific Ring of Fire, this is a land of paradox: fire and rain, lava and
ice, change and resilience, ancient and ephemeral. Move your fin-

ger along a map of the Pacific Northwest, starting at the north-western edge of California. Brush the tall tops of coast redwoods as you move north into Oregon. Follow the dark-green outline of the rainforest as it skirts around the jumbled ancient rocks of the Klamath Mountains and splits into two parallel ranges: the Cascade Mountains and the Coast Range. Plunge your hands into the damp Coast Range forest, dripping with moisture wrung from ocean clouds and sifted through fine needles. Feel your way north along the coast and across the Columbia River until your finger bumps into the Olympic Mountains on the northwest corner of Washington. Cross the Salish Sea to Vancouver Island and sweep across the southwestern edge of British Columbia. Then go back across Puget Sound, follow steep, tree-covered slopes up to Mount Rainier (or *Tahoma,* as some Native people know it), the tallest mountain in the region. Bump south along the sharp edge of volcanoes at the crest of the Cascades and listen to the music of the old names: there is Mount Baker (known as *P-kowitz*), Mount St. Helens (*Loowit*), Mount Hood (*Wy'east*), Mount Adams (*Klicki-tat*), Mount Jefferson (*Seekseekqua*), Three Sisters (*Klah Klahne*), Mount McLoughlin (*M'laiksini*), and Mount Shasta (*Uytaahkoo*).[3] As you travel along the ridgeline, glance over your shoulder at pines and junipers growing on the dry, high desert east of the crest of the Cascades. From Crater Lake (*Giiwas*), move west and back to the foggy Pacific coast. Offshore and out of sight, two tectonic plates squeeze ever more uncomfortably along a seven-hundred-mile sub-duction zone that parallels the western edge of the region.

It's a lively place. You are about to embark on an imagined expedition into this rainforest and its fiery origins, drenching rain, checkered history, and clouded future. As society grapples with how to think about forests—shall we worship them or consume them?—you will see how much is involved in just *being* a forest. The first thing you will notice is the enormous size of the trees. You will drift up to the canopy through masses of ferns and lichens, burrow down into soil through hair-thin threads of fungi, and plunge headlong through a watershed flushed with rain and snowmelt. You will experience the rainforest as it faces a shifting climate and the shifting priorities of a constantly changing society. You will travel across scales from microscopic to planetary, and from moments to millennia, beyond the grid of latitude and longitude, into places only your imagination can fit.

For thousands of years, forests have set the stage for people to tell stories about themselves. The forest might be home to hungry witches or angry, fire-throwing gods, and it could be a refuge for merry men, tricksters, or Sasquatch. The forest of Paul Bunyan extended across the entire continent of North America, where the mythical lumberjack cut and stacked the wilderness as quickly as he could. A forest might be haunted by violence, graced by spirits, or ransacked by greedy opportunists. All forests share a human history of conflict between veneration and consumption. What about the forest itself? What is its story? The forest of fire and rain has its own epic history of disturbance and endurance. It is an *intemperate* rainforest, here at the restless edge of a shifting continent, and this is its story.

When Europeans set foot in North America, most of the continent was covered with woodlands. European settlers whittled away at those woods in their push for new land, and by the time of the American Civil War, the northeastern forests were essentially logged out. Lumberjacks moved on to the forests of the Great Lakes, the South, and finally to the Pacific Northwest, cutting down trees with increasing speed and machinery. By the end of the nineteenth century, the timber industry hit the Pacific Ocean, and soon the endless supply of American timber was in question. A conservation movement emerged in the 1910s, with a scientific approach to rebuilding America's cutover woodlands. New forests were planted, and the old forests of the Pacific Northwest, the nation's last unlogged stands, were to be replaced with efficient timber-producing plantations. Cutting down giant, ancient trees led to another conservation movement in the 1980s, this time focused on old forests in the range of the northern spotted owl and on radical new research that would reimagine the importance of a forest for something beyond lumber.

My husband and I moved to Oregon's Coast Range in the 1970s and built a house *in* and *of* the Douglas-fir forest. We salvaged boards from an old barn, rough-hewn planks cut fifty years earlier from old-growth Douglas-fir trees. We lived in that house for seventeen years, tinkering as you do with a much-loved, never-finished, hand-built house. We were part of a back-to-the-land community of carpenters, loggers, tree-planters, teachers, nurses, and artists. Trained as an ecologist, I first worked as a fish biologist, chasing salmon up and down forest streams. Later, as a science writer, I followed other scientists deeper into the mysteries of those streams and forests. I filled sketchbooks with what I saw along the way.

Much of the research we'll visit in this expedition has been conducted by federal and university scientists, particularly by a cadre of upstart researchers at the H. J. Andrews Experimental Forest in the Cascade Mountains of Oregon, where the scientific study of old-growth forests first took root. Their groundbreaking discoveries challenged America's two-hundred-year-old engine of intensive logging. As a result, the temperate rainforest of western Oregon and Washington is among the most studied ecosystems in the world. This is where the owl wars were fought on behalf of a thousand other species that live within the old-growth forest. This is where federal forest management was turned on its head. This is where I have spent most of my life, writing about the landscape and the people who pursue its mysteries.

As you will see, who owns the forest land makes a big difference in terms of what the forest looks like and how it lives. You, as part of the American public, own 60 percent of Oregon's land and 42 percent of Washington's. The Cascade Mountains form a nearly continuous line of publicly owned federal land from northern California to the Canadian border, including several shoulder-to-shoulder national forests. The Coast Range, by contrast, is made up of a collage of landowners, including several state, tribal, and national forests; a checkerboard of private land alternating with public Bureau of Land Management land; and the Olympic National Park. As seen from the air, the region's mountains are a jigsaw puzzle of clear-cuts, plantations, and wild forests that are naturally generated, messy, and whole. That requires explanation.

A tree plantation is an invention of twentieth-century forestry, a block of trees planted specifically for efficient timber harvest. A wild forest is much more complex. It's like the difference between a cornfield and a prairie. Corn is a crop, intentionally planted as a monoculture to produce one single product. In contrast, a prairie emerges on its own, with a multitude of plants and animals, as a complex, self-perpetuating ecosystem not dependent on humans to add water, fertilizer, or herbicides. It's an important difference, even if the labels fail us. So let's agree to use imaginatively the term "wild forest." Imagine a forest that has been allowed to take root on its own, face its own challenges, grow up and possibly grow old, and regenerate according to its own ways. Let's call this a wild forest. Wild forests grow from biological riches left over from a fire, flood, or some other natural disturbance; cornfields and clear-cuts lack such legacies.[4]

A wild forest is a kaleidoscope of many species, young and old, large and small, living in communities that produce, consume,

and decompose through time. It's tempting to think of a stately old-growth forest as the epitome of this ecosystem, when really it is just one time stamp on a long procession of change. This temperate rainforest is always changing. Death in individual trees bursts with returning life. Upheavals from fire, rain, pests, volcanoes, wind-storms, and landslides have all had a hand in sculpting this magnif-icent forest. The biggest changes in the last century have been from human hands.

This temperate rainforest is among the most timber-productive regions in the world, and productivity has its costs. Although European settlement came relatively late to the Pacific Northwest, the region has weathered relentless logging in a short amount of time. In just a few generations, logging has left large swaths of young plantations covering much of the lower eleva-tions. As the region's yearly timber production hit over fifteen bil-lion board feet in the 1980s (that's enough to build more than one million houses each year), the Douglas-fir region became a battle-ground for protecting the remaining old forests.

Despite all this logging, millions of acres of wild rainforest still stand throughout the region, mostly on U.S. Forest Service land.[5] Occasionally, we will wander into a tree plantation, but most of the forests we will explore are naturally generated, intact, and wild. Within these forests are stands, or patches where a group of trees has established at the same time and grown up together. Old forests are sometimes referred to as old growth, a term apparently preferred by old foresters. There are other terms—"habitat," "com-munity," "seral stage," "landscape"—that bring their own knives to the table for cutting the forest up into bite-size chunks for manage-ment or research. We'll walk fearlessly through them all.

The tallest Douglas-fir reliably measured was 393 feet tall. Others cut earlier may have been even taller.

In this imagined field trip, we will explore what it means to be a forest in a rapidly shifting world. Our exploration grows in complexity from the initial gee-whiz moment you first enter the rainforest to the concluding chapters, when you recognize yourself as an agent of change in the world. It is critically important to understand how this ecosystem works. The reality of climate change has accelerated environmental action around the globe. Taking action is good, but not every action will result in positive change. Reengineering the planet's atmosphere is a dangerous way to avoid the decision to stop burning fossil fuels. Similarly, exploiting the public fear of wildfire has reopened the door to industrial logging and suspended environmental laws that would protect mature and old forests, which hold the most fire-resistant trees in the forest.

In the United States, the policies that implement environmental laws are being revised for the first time in a generation.[6] The public is being asked to comment on sweeping changes to federal forest management, as well as regional water shortages, carbon emissions, and many other issues. Understanding how complex natural systems *actually* work is essential.

For many people, this lush Pacific Northwest rainforest is an unchanging backdrop to modern human lives; few people give it much thought beyond its magnificent scenery. Forests in the Amazon or the California redwoods grab more attention. And so it is too often we overlook nearby wild places. Each generation manages to save a fraction of its natural environment inheritance. At that rate, it doesn't take many generations to whittle down a great forest into plantations of spindly telephone poles. Nobody decides we should cut the last thousand-year-old tree. It just happens. It's the slippery slope, the boiling frog. It's happening now, and no one notices because, after all, it's just one tree.

Luckily, we are not yet down to our very last tree. Old forests can still be found, despite rapacious logging in the past and unknowable challenges in the future. When I first moved to the Coast Range, my elderly neighbors told me stories of the enormous trees they saw as children in the early twentieth century. Perhaps my grandchildren will live to see trees nearly that big at the turn of the twenty-second century. We sit between these two generations, with the fate of the world in our hands.

Unlike stories that mourn the decline and fall of ecosystems, this book is not a eulogy. It is an invitation to explore, pay attention, and understand what is at stake in the world. Because this forest of fire and rain is so big and so tumultuous, it is a good place to consider life on a rapidly changing planet far beyond human control. At this particular moment in history, when the earth is shifting all around us, understanding what it means to be a forest might help us understand what it means to be human.

2

You Enter a Land of Big, Old Trees

The forests of the Douglas-fir region are nearly without rival in the world. Their uniqueness and features . . . are, however, rarely fully appreciated.

—JERRY FRANKLIN, forest ecologist

There is more to a forest than trees, but nothing is more obvious. So, we begin with trees. When you enter an old Douglas-fir forest, you are staring at the ankles of giants. The light is emerald green; the air is cool. It is quiet. The forest rises above a lumpy carpet thick with fallen logs, mosses, and ferns. It reaches up twenty stories in a continuous scaffold of layered branches to treetops far out of sight. Massive trunks stand like columns for more than one hundred feet before they begin to branch. The trees are huge, and some are very old. And all these big trees are conifers.

When it comes to conifers (cone-bearing trees), the Pacific Northwest has the tallest species in the world. We have the tallest spruce (Sitka); the tallest cedar (western redcedar); the tallest hemlock (western); the tallest true fir (noble); and the tallest pine (sugar). No other place in the world can equal the tree species in this region for their size and longevity. There are sequoias that grow bigger and bristlecone pines that live longer, but nowhere else are there so many different species of big, long-lived conifers. And here, if you point to a conifer west of the Cascade crest, there's a better than even chance it's a Douglas-fir.

Hikers follow an old-growth trail in the Gifford Pinchot National Forest.

The tree that lends its name to the Douglas-fir region is the coastal variety, *Pseudotsuga menziesii* var. *menziesii,* the second-tallest tree species in the world. As a species, Douglas-fir has a deep bench of genetic options to take advantage of a wide range of environmental conditions. In the southern part of the region, coastal Douglas-fir grows among dry-loving oak and pine; near the Pacific Ocean, it mingles with perpetually damp Sitka spruce. Its shorter, stockier cousin (*P. menziesii* var. *glauca*) prefers drier climates inland as far as Colorado and Mexico, far beyond the scope of our exploration. You can see plantations of coastal Douglas-firs as far away as France and Chile, where they have been imported as a quick solution to deforestation. This tree has habits that endear it to foresters around the world. On open land, Douglas-fir seedlings burst forth with sun-loving exuberance, and Douglas-firs can continue to dominate almost every stage of forest development for the next six hundred years or more. After several centuries, and with no stand-replacing disturbance, the dominance of Douglas-firs will even-

The Lewis River winds through stands of Douglas-fir and western hemlock in the Gifford Pinchot National Forest.

tually give way to more shade-loving western hemlocks. This alpha-omega relationship defines much of this part of the rainforest as the Douglas-fir–western hemlock vegetation zone.

A lot of what we know about the Douglas-fir region began with forest ecologist Jerry Franklin, one of the first scientists to study this ecosystem in depth. "Douglas-fir is *the* reason these forests are distinctive. It is absolutely unique among tree species," he tells me.[1] I first met Jerry while on a field trip in the Washington Cascades in the 1990s. He was introducing a group of graduate students to a stand of old Douglas-firs, when he arched his back and gestured toward the canopy high above us. "Look at these trees!" Franklin exclaimed. We followed his gaze. "They each have such individual character!" Huge trunks towered out of sight. Far overhead, aerial gardens of ferns, huckleberries, and even young hemlocks had sprouted on heavy limbs, burying their roots into deep mats of composting moss. "There are entire ecosystems up there, out of sight!" he said, still leaning back and gesturing in wonder.[2]

In the classic publication that defined the Douglas-fir region, Franklin wrote, "Most temperate forest regions in the world (Asia, Europe, and eastern North America) have natural forests in which deciduous hardwoods dominate, with conifers concentrated on sites with harsh environments or occurring mainly as pioneer species. In the Douglas-fir region, roles are reversed."[3] If you go back three hundred million years, you see a world full of conifers. With flowering plants and grasses yet to evolve, conifers formed vast forests with three-hundred-foot trees towering over giant ferns and horsetails that sheltered millipedes as long as rattlesnakes and dragonflies as large as hawks. This was the age of giants—dinosaurs and conifers. As the climate grew steadily drier, conifers became isolated in moist clumps. With the rapid rise of flowering plants, most conifer species were pushed out to the edges of shifting continents, or to extinction. More than twenty thousand conifer species covered the planet one hundred million years ago, compared with fewer than six hundred living today. The rise of flowering plants and the fall of conifers illustrate how changing climates can prune off entire branches from the Tree of Life.[4]

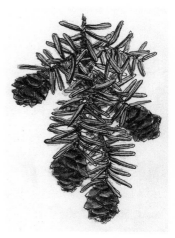

Of course, this patch of the Pacific Northwest did not evolve as a uniform mass of big, old Douglas-fir trees. Across gradients of time and distance, forests change with climate and circumstances. Between the forested Cascades and Coast Range are distinctive, non-forested lowlands— the Willamette Valley in Oregon, the sculpted coastlines of Puget Sound in Washington, and the Salish Sea in southern British Columbia. On the fog-drenched coast, Sitka spruce holds its own with a fringe of smaller shore pine. Four species of conifer, all referred to as cedars, grace the rainforest. None are true cedars. Three are in the cypress family: Alaska yellow-cedars tend toward cooler, wetter areas to the north; incense-cedars tend toward warmer, drier areas to the south; and Port-Orford cedars inhabit a small strip of southern coastline where coast redwoods transition to Sitka spruce. Western redcedars (related to arborvitae) live throughout the region.

Western hemlock needles come in a variety of lengths and point in all directions, giving the tree its species name, *heterophylla* (varied leaves).

The Pacific yew carries its seed in a bright-red aril rather than a cone, the only flashiness of this small, wizened tree.

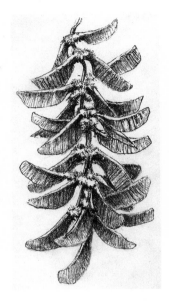

Western hemlock is ubiquitous in the rainforest, yet is often, and literally, overshadowed by Douglas-fir. The tallest hemlock species in the world, these trees start as seedlings balanced on mossy stumps or tiptoeing in a line across a fallen log, where their tiny seeds find a rich, moist substrate for putting down roots. Western hemlocks can remain small, suspended in this bonsai state, for a century or more, waiting for a gap to open in the canopy and make room for towering growth.

The Pacific yew is another patient, slow-to-grow conifer. Small in stature and concealed under thick robes of moss, this inconspicuous tree made headlines when its bark was

found to contain a chemical compound that is effective in treating cancer. A synthetic compound since developed has saved wild Pacific yews, some more than one thousand years old, from being stripped of bark and left to die.

Although this is the land of conifers, two deciduous trees are champions in their own right. Big-leaf maple displays many superlatives, besides its dinner plate–size leaves. It is the largest maple in North America and a champion of regeneration, with a prodigious

Profusions of winged samaras carry the seeds of big-leaf maple, the largest maple in North America.

Pendulous catkins and cone-like strobiles decorate red alder, a ubiquitous tree along low-lying rivers.

ability to sprout from broken stems. Tucked into damp spots on lower slopes, maples have generous limbs that hold thick blankets of moss, lichens, and ferns. Big-leaf maples are every small person's favorite climbing tree.

Less flashy but no less important, red alder is common along Coast Range rivers. It is one of the largest alder species in the world, with marble-like bark splotched with thick colonies of silver-white lichen. But it's the red inner bark that gives red alder its name, as well as its usefulness as a natural dye and a pain reliever (it contains a chemical similar to aspirin). Fast-growing and able to deliver nitrogen from the air, red alders boost forest fertility.

There are plant communities in this forest beyond the reach of big trees. Mountain meadows and rock gardens show up in spots where trees struggle to take root—on thin soils below ridgetops or at the edge of landslides. Here in late spring, snow-white topknots of beargrass bobble like something from Dr. Seuss. The meadows themselves are crawling with colorful, Seussian diversity. Blue lark-

---

Sitka spruce, the world's largest spruce, towers over the coastal fog belt within a few miles of the Pacific Ocean.

spur, pink penstemon, yellow stonecrop, and many more flowers
in riotous color attract pollinating insects and insect-eating birds.
Coastal meadows hunker down on high, hard bedrock promonto-
ries overlooking the Pacific. On these fog-swept headlands, Sitka
spruce and western hemlocks tuck into protected coves, along with
some of the region's rarest plants and butterflies. Before European
settlement, meadows on both mountain slopes and coastal headlands were routinely managed with low-intensity fires to encourage
a flush of first foods such as bracken fern, camas, and beargrass.

Little of this biodiversity held much interest to early loggers beyond Douglas-fir. While no reliable records exist on the age or size
of the first giant trees harvested here in the late nineteenth century, old photographs and newspaper accounts suggest that the
biggest, oldest Douglas-firs may have been over four hundred feet
tall and nearly one thousand years old, possibly the tallest tree species on the planet.[5]
     Why are these trees so big? In part, it's the Goldilocks climate
of the Pacific Northwest: not too hot, not too cold, with just the
right amount of moisture. Hurricane-force winds are relatively rare
here, something that would otherwise limit the height of the most
ambitious trees. Conifers can grow throughout much of the year in
this mild climate; a Douglas-fir absorbs half its annual net carbon
between October and May, when deciduous trees are mostly leaf-
less. And the conifers in this region continue to grow for centuries,
long after trees elsewhere have essentially maxed out. This moderate climate also affects fire frequency. Too much fire might push
this region to resemble the pine forests of the interior West. Too
little fire, and it might become like the moss-shrouded cedar forests of southeast Alaska.
     Could this Goldilocks climate favor other tree species as
well? That was a question posed by a Yale Forest School graduate named Thornton T. Munger, one of the first scientists to put

his mind toward growing Douglas-fir. Munger arrived in the Pacific Northwest in 1908 when the U.S. Forest Service was just taking root. As a one-man research unit in the regional Forest Service office, Munger focused much of his study in the Wind River area of the southern Washington Cascades. There he found a patchwork landscape of burned forest—some very recent and some hundreds of years old—to compare how large and how fast Douglas-fir could grow. In 1912, Munger hired local loggers to clear eleven acres of wild forest for an arboretum where he could test how other tree species would grow in the western Cascades. Just clearing out those massive stumps must have taken a Herculean effort. Munger's team tested seeds from 152 varieties of conifer from all over the world. Some seeds failed to germinate; others grew well enough until summer drought or winter snow took a toll. Clearly, this Goldilocks climate wasn't everybody's idea of paradise. The Wind River Arboretum experiments lasted seventy-five years, and today the place is overgrown with vigorous native trees obscuring the bones of the non-native trees. By far the most conspicuous part of the arboretum is the towering perimeter of Douglas-firs that have grown over one hundred feet tall during the century in which the imported trees shriveled and died.[6] In contrast, Douglas-firs have been successfully exported to Europe, South America, and New Zealand, where they outgrow most native conifers.

Throughout the Pacific Northwest, you can see individual Douglas-fir trees that look quite different from one another growing in the same location. Munger saw this, too, and wondered how adaptable these variations might be. To find out, he collected seeds from several places across the region and planted them in different locations, launching one of the first forest genetics studies in North America. Generations of researchers have monitored the fate of these Douglas-firs as the trees matured. A century of data show Douglas-fir trees can adapt to new locations across a remarkably wide temperature range of two degrees Celsius cooler or warmer and still retain long-term survival and productivity. It's interesting to note that Munger's 1912 study led to the two-degree-Celsius mark as the limit for successful Douglas-fir adaptability. Eighty years later, the international community of climate scientists identified two degrees Celsius as the upper limit of global warming before we see life on Earth struggling to adapt.[7] Perhaps Douglas-firs' ample genetic diversity will help this rainforest adapt to the changing climate.

Since Munger's experiments, we've learned it takes more than a mild climate to grow big trees: you need tree species with the genetic potential for bigness. Douglas-firs make the most of low winter light, with steep, cone-shaped crowns that absorb slanting rays. Their evergreen needles are wrapped in weathertight, waxy coats with very little freezable material inside. Breathing pores are tucked into a narrow groove on the needle's underside to keep moisture in. A mid-size Douglas-fir can have sixty million needles, and it can hold on to those needles for seven years, reducing the energy needed to push out a headful of new foliage each spring.

Most of the longest-lived conifers of the Pacific Northwest grow big relatively quickly, with adaptations to protect themselves for the long haul. Big Douglas-firs, for example, shed lower branches to deter ground fires from laddering up to the crown, and they develop corky, fire-resistant bark up to eight inches thick to rebuff all but the hottest fire. They carry a medicine chest of biochemicals that resist decay and harbor a friendly force of fungi to help fight off pathogens. And they store a lot of water in their sapwood that helps moderate periods of drought. With such built-in defenses, these trees can live a long time, with centuries to develop a complex architecture that creates all sorts of habitats and natural communities aboveground and belowground.

There are also structural tricks that keep these tall trees towering, both below the ground and above. A Douglas-fir seedling quickly grows an initial taproot to anchor it to the ground. After a few years, lateral roots develop to stabilize the tree. With increasing age, the tree develops a deep-reaching system of perpendicular and slanting roots that help ensure stability against wind and shifting ground. A Douglas-fir's roots readily fuse with the roots of others, blurring the boundaries between individual trees and between tree species. Connected roots hold trees to the ground, and they can keep a tree stump alive long enough for it to scab over the cut with a new layer of bark above and an uninterrupted network below.

The trunk of a Douglas-fir is layered like a radial tire, spiraling like a Slinky at the core and straightening toward the outside of the trunk as the tree matures. This spiral allows the young sapling to bend in all directions with the wind. As the tree grows in girth, it increases rigidity in the trunk to provide the support necessary to hold up tons of wood. Cells toward the center of the trunk harden with a strong, glue-like substance called lignin. These cells are no longer involved in the living business of transporting nutrients between needles and roots. That business belongs to a thin band of cells just beneath the tree's bark, called the vascular cambium.

The vascular cambium produces the tree's essential plumbing—phloem to the outside and xylem to the inside. Phloem conducts sugars from the needles down; xylem conducts water and minerals from the roots up. As the tree grows, the vascular cambium builds a ring of xylem cells that form sapwood, while the older xylem cells, infused with lignin, harden into heartwood. Meanwhile, new phloem cells push older phloem cells out to where they turn corky and eventually become tree bark. The thick furrows of Douglas-fir bark, the puzzle pieces of ponderosa pine bark, and the stringy strips of western redcedar bark are all the work of old phloem cells.

It's not easy being tall. Just staying hydrated is a challenge. If you've lived in a high-rise, you know that pumping water up twenty stories leads to less water pressure on the top floors. So it is with tall trees. Plants transport water from the roots to leaves by capillary action through a network of pitted tubes and cavities in the xylem. "Some materials, like sand, have no strength when they are pulled in two directions, but water is surprisingly strong in tension," explains Barb Lachenbruch, a forest engineer with a keen understanding of how trees grow. "Plants take advantage of this, just as you do when you suck water up a straw. Because the water is continuous through the roots, trunk, branches, and leaves, evaporation from a leaf into the drier atmosphere pulls water upward." Very tall trees must transport a lot of water up to their sprawling heights, so their xylem cells grow broad and long toward the outer edge of the tree. Such extended cells are particularly vulnerable to forming freeze-and-thaw bubbles that can clog the works. That's called embolism, and it can be as life-threatening to trees as it is to people. Trees that grow where winters are cold enough to freeze their tubes can't afford the plumbing it takes to grow very tall.[8]

Drought can do something similar. Warm, dry air tends to pull moisture out of tree leaves; such evapotranspiration stretches the water column inside the xylem like a rubber band. When the air gets very dry, the stretched-out water column can pull apart, pulling in air bubbles that block further flow. To counteract this, on a hot, dry day, trees close the pores on their needles to stop water from escaping. Closing pores also stops photosynthesis, essentially shutting down the tree's engine for energy and growth. Most trees that live where summers are hot and dry can't afford the moisture it takes to grow very tall. By calculating the complex relationship between efficient water transport and the risk of xylem embolism, Lachenbruch and colleagues have determined that the absolute height of a Douglas-fir tree may be limited to 430 feet, the height of the Great Pyramid of Giza.[9]

Standing among old Douglas-firs, you see their massive trunks rise straight and branchless far into the sky. However, centuries earlier, when these old giants were just young sprouts, their branches whorled around their lanky young trunks like layers of petticoats. Like many of their pine cousins, young Douglas-firs produce new branches from buds that surround the tip of the so-called terminal leader. One bud will become next year's leader, and the rest of the buds rally around their leader in a circle of perpendicular branches that ring the growing trunk. The tree develops a single tall, straight trunk and layers of branches that can fold like an umbrella under the weight of heavy snow without breaking.

Eventually, lower branches become heavily shaded and drop off the tree, leaving just a small scar. Over decades the scar will be masked by expanding bark, but the memory of the phantom branch remains. If a gap of sunlight opens centuries later, a new branch might grow from the old scar. This new limb, called an epicormic branch, ignores the old rules of perpendicular growth and can develop more like a separate parallel trunk, sometimes massive. Over time, a Douglas-fir outgrows its simple, whorl-based petticoat to become a unique individual with unruly branches and parallel trunks replacing its more youthful tapered profile.

Although the first people in the Pacific Northwest revered western redcedar for their longhouses, canoes, fabrics, and sculpted poles, Europeans who arrived in the 1800s were taken by Douglas-firs. David Douglas, the Scotsman for whom the tree was named, was the rare early European explorer who was actually trained as a botanist. During his three visits to the Pacific Northwest between 1824 to 1834, he described hundreds of plants for science, including the Douglas-fir. Clearly impressed by the rainforest's diversity of conifers (he called them all pines), he wrote to the Royal Horticultural Society, "You will begin to think I manufacture pines at my pleasure." Eventually, Douglas sent samples of the tree to London, where the British botanists labeled it Douglas pine. Today, more than eighty species of plant and animal have "*douglasii*" in their scientific names. (For a rare, first-person tour of the Douglas-fir forest two hundred years ago, and the foibles of an eager, young, tea-sipping explorer, read Douglas's digitized journals.)[10]

But Douglas was not the first European to notice the Douglas-fir. Archibald Menzies first described it during a 1792 exploration of coastal British Columbia. Meriwether Lewis had plenty of time to

study it during the long, wet winter of 1806, referring to the tree in his journals as "Fir No. 5." Early loggers called it red pine; John Muir called it Douglas spruce. But the tree is not a pine, spruce, or true fir. It is in a genus all its own, eventually christened in 1953 as *Pseudotsuga* (false hemlock) *menziesii,* in honor of Archibald Menzies, who introduced it to the world but had not bothered to name it.

It was not fir but *fur* that brought the Hudson's Bay Company to the Pacific Northwest in 1821, initially to hunt beaver and sea otter. Fur trapping was seasonal work, so their exploitive business model quickly expanded to include shipping salmon to London and wood to Honolulu. With the gold rush boom in California, timber became a pillar of the nascent Northwest economy.

When Europeans set foot in North America, the land they encountered was heavily forested. Settlers used wood for fuel, building material, railways, and fences, and they cut their way across the continent to the edge of the Pacific. After 1846, when the forty-ninth parallel was established as the border between the United States and British-held Canada, the British Hudson's Bay Company sold its southern land holdings to American entrepreneurs eager to claim their piece of the Northwest timber empire. By the mid-1850s, there were over two dozen lumber mills on Puget Sound, cutting

---

Western redcedars can live for more than a thousand years, and their downed wood can last almost that long.

down giant Douglas-firs within easy reach along lowland rivers. The American timber industry quickly expanded into the lower Columbia and Willamette river valleys and the southern Oregon coast at Coos Bay. As timber fell, the timber industry pushed inland and upslope. By the beginning of the twentieth century, logging companies were building massive log bridges across steep canyons to carry giant sawlogs out of the hills and to the mills.

Timber harvest was laborious, particularly on rugged upland terrain. Cutting down a single giant Douglas-fir could take days. At that rate, much of the original rainforest stood uncut until chainsaws, trucks, and tractors came to the woods in the 1940s. Today, you can still see huge stumps of trees that were cut nearly a century ago with double-bit axes and long crosscut saws. The stumps are punctured with slit-eyed holes that held springboards where loggers stood to make their cuts above the flare of roots. My elderly neighbors saw some of these enormous trees fall. A generation later, I witnessed chainsaws cutting down what was left at a

---

This burl is a benign growth bulging from a perfectly healthy Alaska yellow-cedar. Chainsaw-wielding thieves amputate these burls and kill the tree in the process.

frantic pace. By the 1980s, three-quarters of the original forests of western Oregon and Washington had been cut down, a testament to unstoppable industrial efficiency.

Only photographs remain of the oldest Douglas-fir giants, pictured with lumberjacks as small as Jack at the base of his giant beanstalk. Early on, timber tycoons recognized the immense value of Douglas-fir lumber: strong enough to build a bridge and fine-grained enough to fashion window trim. Douglas-fir built the colonial settlements in Hawai'i. It built San Francisco during the gold rush and rebuilt San Francisco after its 1906 earthquake. Frederick Weyerhaeuser considered Douglas-fir to be his money tree. It helped fuel the postwar building boom across the United States, and today engineered mass timbers of Douglas-fir have begun to replace steel in high-rise buildings. It also makes a lovely Christmas tree.

My husband and I built our house in the Coast Range with old-growth Douglas-fir boards we recycled from a ramshackle barn, and roofed it with western redcedar we pulled out of a log-jam on Big Elk Creek. With a wedge-shaped froe and a wooden mallet, I snapped off fat slices of fragrant, straight-grained shakes from that cedar, splitting enough shingles to roof the house just ahead of the winter rains. Years later, that beautiful cedar roof caught fire, but that's another story.

For generations of people, this temperate rainforest has provided homes, jobs, resources, PhDs, epiphanies, and solace. Over time, people have valued this forest for building material or for biodiversity, as a source of clean water or a source of industrial wealth. Now the forest is valued for carbon sequestration and feared for the perceived flammability of that carbon. When we focus on one value alone, we risk overlooking the complicated picture of everything tangled together through time. This rainforest has evolved in a landscape that slides, burns, explodes, and rots, a place that is constantly changing.

3

You Stand on a
Rollicking Foundation

Geology is destiny.

—GORDON GRANT, research hydrologist

When you stand at the top of a riverbank, you are at the edge of two worlds, with a geology lesson in between. There's also a lesson in gravity, as the ferns beneath your feet start to give way to an irresistible tug toward the stream below you. Unable to keep your footing, you begin to slide with the ferns. You grab a banister of tree root. Now you're suspended on a vertical bank that was carved by the hydraulic power that pushes the fastest-moving water against the outside of a curving river channel. You are a mass wasting event, and you're about to get wet.

Hold that thought. There aren't many places in this moss-draped place where you can see the rock that holds up the forest, so take a moment to look around. A timeline of geologic history is right in front of your nose as you hang suspended in a geologic instant. You are slipping down through a few decades of accumulated gravels, and past a few million years of volcanic deposits. You are eye-to-eye with compacted glacial till washed from glaciers that melted ten thousand years ago. It's a rare look at the foundation of the forest.

Grip that root. You are hanging by a thread on a planet that constantly rumbles, erupts, cracks, and creeps with processes that

Lower Toketee Falls tumbles eighty feet down a wall of columnar basalt in the Umpqua National Forest.

both support life and destroy life. Throughout time, this place has shifted its composition in response to external upheavals. Change can happen at any moment, even to you.

You can let go now.

The massive forests of the Douglas-fir region, growing in one of the most productive regions on Earth, are atop one of the most tectonically active places on Earth. Let's begin by bumping along on the gigantic North American plate, one of many tectonic plates that form Earth's crust. Plates knock around each other like rafts on a crowded swimming pool. We are floating atop a mantle of hot, churning, gooey rock, pushed along a geologic conveyor belt from the point where new seafloor oozes from a mid-ocean ridge to where that seafloor eventually gets sucked back into the mantle. Oceanic plates are heavy; continental plates are lighter, more buoyant. So when they collide, continental plates ride over the top of oceanic plates, pushing the heavier raft down into the hot mantle at the subduction zone.

That's what is happening as our edge of the continental North American plate bumps over the top of the oceanic Juan de Fuca plate, pushing it down into a pressure cooker. Rocks and water get sucked in, squeezed down, and heated up; volcanoes pop up like an adolescence of pimples. (Currently, these volcanoes are strung on a seven-hundred-mile volcanic chain along the crest of the Cascade Mountains.) Where the oceanic plate is too thick to slide neatly under the continental plate, oceanic rocks are scraped off at the edge of the continent. These scrapings form the Coast Range. This bank of oceanic basalt (known as Siletzia), covered by deep layers of eroded sediments, is lifting up as the subducting Juan de Fuca plate noses underneath it.[1]

This is more than simply nosing around. Seventy miles offshore from Oregon and Washington lies a megathrust fault of enormous consequence. Unlike the side-to-side grinding of the San Andreas Fault, whose frequent earthquakes release tectonic ten-

sion, the Cascadia sub-
duction zone has been
building up tension
for hundreds of years.
Wedged between oce-
anic plates to the west
and a continental plate
to the east, the subduc-
tion zone appears to be
stuck in an increasingly
uncomfortable position,
locked by friction while
tectonic stress increases
below and human anxi-
ety increases above.[2]

Megathrust earth-
quakes are the most
powerful earthquakes
known on Earth. They
can exceed magnitude
9.0, which releases one
million times more en-
ergy than a magni-
tude-5.0 earthquake.
It's not a question of *if*
such a quake will occur,
but rather *when,* and
when it does, the Cas-
cadia subduction zone

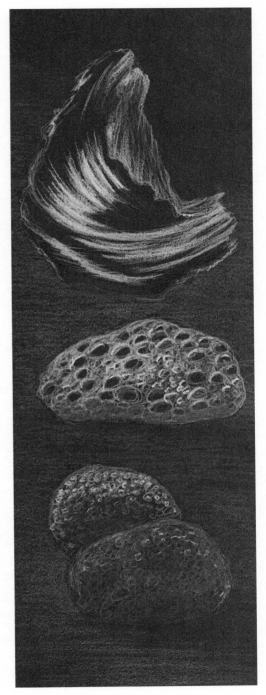

Obsidian, pumice, and
andesite rocks sketched at
Mount St. Helens National
Volcanic Monument.

could launch enormous tsunamis with incalculable destruction to cities, power grids, forests, and human lives. The Cascadia subduction zone has produced almost forty magnitude-8.0 or larger earthquakes during the past ten thousand years, on average once every 234 years.[3] By that measure, Oregon and Washington are nearly one hundred years overdue.

When settlers followed the Oregon Trail, they had no way to even imagine the geologic forces inching toward the devastation of this paradise they thought of as their manifest destiny. It wasn't until 1970 that geologists had any clue this Cascadia subduction zone existed. And not until 2011, when a magnitude-9.0 earthquake hit the Tohoku region of Japan, did we have a modern example of what could happen to all that tectonic energy pent-up in a subduction zone. The Tohoku earthquake and subsequent tsunami triggered the Fukushima power plant meltdown, killed more

---

Stumps of an ancient Sitka spruce forest emerge from the surf where a subduction zone earthquake dropped the land and this forest in 1700.

than eighteen thousand people, and cost more than $220 billion. An earthquake along the Cascadia subduction zone could be this destructive. And because a Cascadia quake will be centered off-shore, it will trigger a massive tsunami that will devastate the Pacific Northwest coast and radiate all the way to Japan.[4]

Apparently, something like that happened more than three hundred years ago, when a very large megathrust earthquake shook the coast of present-day Oregon and Washington. Native people in the region pass down stories of shaking ground and a saltwater flood: "The earth sank into the water," they said, and "everything was lost and gone."[5] The earthquake caused coastal bluffs to suddenly collapse, taking whole chunks of forest with them. Trees still rooted to the ground had suddenly dropped to sea level and were eventually buried in sand. You can walk among the remains of one of these sunken forests at the small Oregon beach town of Neskowin, where ghostly tree stumps encrusted in barnacles emerge from the surf at each low tide.

When scientists compared tree rings of these tide-splashed ghost forests with tree rings of living trees still standing on higher ground nearby, they were able to narrow the tsunami event to around the year 1700. Further investigation using old Japanese records of earthquakes and tsunamis revealed that the last big earthquake to rattle from the Cascadia subduction zone was on January 26, 1700, at about nine o'clock at night. This scientific detective work is a worrisome data point for people living on the edge of the Pacific Northwest. Unraveling the relative frequency of these earthquakes, seismologists estimate that in the next fifty years, the odds of a Big One (a partial slip of the Cascadia subduction zone) are one-in-three, and the odds of a Really Big One (a full release of the megathrust fault) are one-in-ten.[6]

When people here in the Pacific Northwest aren't fretting about the Big One, they are fretting about the rain. If you think it rains all the time in this rainforest, you're half right. It rains half the time, and the other half not so much. North to south, annual precipitation is on a continuum from gusher to trickle. In an average year, southern Oregon receives eighteen inches of precipitation, while the western Olympic Mountains in Washington receive twelve *feet*. And the contrast is equally stark east to west. The Olympic Mountains, Oregon's Coast Range, and the Cascade Mountains all create formidable barriers between the wet, ocean-facing west side and its drier, continent-facing east side. It's a difference you sense within a few miles on either side of the Cascade crest, as moist, Westside Douglas-firs give way to dry, Eastside ponderosa pines. At the Cascade crest, more than one hundred inches of precipitation can fall in one year; that average drops precipitously to less than twelve inches on the eastern flank of the Cascades, only forty miles to the east. And it's not just rain: at higher elevations, the snowpack can be impressively deep. The *world* record for annual snowfall belongs to Mount Baker in the southern Washington Cascades, where forty-five feet of snow fell during the winter of 1999.

What happens to all that water? Let's start near the top of the Oregon Cascades, at the summit of the McKenzie Pass in Oregon. Here, the Dee Wright Observatory, built by the Civilian Conservation Corps in 1935, rises out of a jumble of volcanic rocks left from an eruption about twenty-six hundred years ago. The view is spectacular. Flows of hardened lava extend to the horizon, ringed by snow-peaked volcanoes. What you don't see from this bird's-eye view is flowing surface water. In this landscape, melted snow and rainwater disappear underground. The water seeps down then flows for miles between buried layers of lava rock, until it eventually gushes up to the surface in springs far down the slope. The water that bubbles from these springs today fell as precipitation years or decades ago and emerges clear, ice cold, and sapphire blue. A mother lode of stored water is hidden in the deep, fractured rocks beneath the High Cascades. Geohydrologist Gordon Grant predicts that with climate change, the most valuable product from na-

tional forests might not be timber, but water—clean, abundant water. And because of their particular water-holding geology, forests
in the Cascade Mountains are the Forest Service's top producers of
this vital resource.[7]

Geologic cataclysms have left other marks on the Douglas-fir region, including immense floods of basalt lava that gushed across
the Columbia Basin beginning about sixteen million years ago.
The Columbia River flood basalts boiled out of huge cracks along
the Oregon–Idaho border in gigantic floods of lava across forty-
one thousand square miles and more than a mile deep. If similar
flood basalts were to burst forth today, they would bury all but the
tip of Mount Rainier. (At 14,411 feet, Rainier is the tallest peak in
the Cascade Mountains.) For three million years, a series of more
than three hundred basalt flows built coastal headlands and stair-
step layers of basalt columns, iconic features of today's Pacific
Northwest landscape.

In between eruption cycles, the lava-encrusted landscape
slowly reestablished forests. Subsequent flood basalts buried these
emergent woodlands and turned the trees to stone. Over time, petrified forests were buried by more and more lava until, millions of
years later, the basalt landscape was rearranged several times by a
different sort of flood: Pleistocene floods of surging ice, rocks, and
meltwater. At least forty times during the last Ice Age, between
eighteen thousand and twelve thousand years ago, ice dams grew
large enough to hold back enormous glacial lakes. And when the
dams burst, surging walls of water hundreds of feet deep scoured
the Columbia Basin and flooded hundreds of miles with enough
hydraulic power to scatter house-size chunks of basalt columns and
deposit sediment up to a half-mile deep in the Willamette Valley.[8]

Along the Columbia River near Vantage, Washington, a log-
jam of petrified wood marks the spot where a forest grew 15.5 million years ago. Fire and rain were at work even then, as the living

trees were buried in an eruption of volcanic ash that gradually replaced the wood cells with dissolved quartz minerals. Subsequent basalt flows smothered the petrified forest in deep layers of lava where the stone trees lay encased for millions of years. Then, fifteen thousand years ago, the last of the Ice Age floods ripped down through lava rock to where the petrified forest lay buried for fifteen million years. Walking among this stone forest today, you see petrified logs from more than forty tree species; the most abundant is Douglas-fir.

Several times during the last two million years, continental ice reached far enough south to bury the Puget Sound area in thousands of feet of ice, sculpting the inland coast into myriad islands. The region's last ice sheet, the Cordilleran, reached its farthest extent about twenty thousand years ago, when the Vashon glacier covered present-day Seattle, Tacoma, and Olympia. After the last cycle of ice and flood, today's Douglas-fir

Layers of soil in the Gifford Pinchot National Forest show the region's history of repeated volcanic eruptions.

forests established about six thousand years ago—a mere handful
of Douglas-fir generations.

The region's long history of volcanic eruptions and ground-up
glacial sediments has created some of the most productive soils in the
world. As Jerry Franklin says, Cascadia soil falls from the sky. Up to
20 percent of the soil in Washington's southern Cascades is weathered volcanic glass.[9] Washington's state soil, Tokul, flew in as volcanic
ash and sediments scoured from glaciers; Oregon's state soil, Jory, is
the result of eroded flood basalts. Both soils grow exceptional forests.

Volcanoes are the region's most prominent geologic feature. Along
the crest of the Cascade Mountains, twenty major volcanoes and
nearly four thousand volcanic vents punctuate the Cascade Volcanic Arc, from Mount Lassen in northern California to Mount
Garibaldi in southern British Columbia. The eruption of Mount
Mazama, sixty-eight hundred years ago, scattered ash across most
of Oregon and left a gaping hole in the Cascades called Crater
Lake, the deepest lake in North America. The beautiful, snow-
capped peaks that grace the skylines of Seattle and Portland are
considered among the world's most potentially dangerous volcanoes. Mount Rainier, for example, has been dozing for five hundred years, yet its massive size, frequent earthquakes, extensive ice
fields, and deep hydrothermal activity could send deadly mudflows
into a region where millions of people now live.

Erupting volcanoes, floods of ice and lava, and a jittery offshore subduction zone remind us that this temperate rainforest is a
spirited place, beyond our imagination and out of our control. That
spirit might have remained a passing thought until 1980, when the
eruption of Mount St. Helens shook up the Pacific Northwest and
rattled the thinking of a handful of scientists. One of those scientists was U.S. Forest Service geologist Fred Swanson.

A few years earlier, Swanson had been scrambling through
dense forest understories in the western Cascades attempting to

map landforms beneath the H. J. Andrews Experimental Forest near Blue River, Oregon. "You couldn't see much of any landforms, because of all the damn trees," he later laughed. Hidden beneath the mossy layers of fallen logs, Swanson found overlapping lava flows cut by streams and layered in landslides. He began piecing together a bio-geo-chemical story of a volcano-born landscape, shaped by fire and floods, blanketed in trees and moss. He saw evidence of move-

before the eruption

8:32:33 A.M.

8:32:43 A.M.

ment beneath his feet. Trees leaned downhill, sliding slowly down an unstable foundation of broken rock and buckets of rain. These slow landslides move imperceptibly to us, but not to a tree that is trying to balance on a slumping hillslope for hundreds of years. Change, Swanson realized, is underfoot.[10]

It took less than one minute for an earthquake to trigger Mount St. Helens's eruption on May 18, 1980. (Drawings based on photo by Gary Rosenquist, USGS)

8:33:00 A.M.

By the spring of 1980, as we hunched over our Apple II computers wondering what to do with all those sixty-four kilobytes, we also wondered when Mount St. Helens would erupt. We looked forward to each update on the Cascade volcano's baby bump, as if St. Helens were about to deliver a bouncing newborn with heart-shaped clouds floating above its perfect triangular peak. Seismographs predicted something far more ominous.

On a sunny May morning, after weeks of small tremors, a magnitude-5.1 earthquake shook loose an enormous eruption on Mount St. Helens. A ten-megaton explosion collapsed the entire north side of the mountain. With the lid off the pressure cooker, hot water beneath the mountain blasted out in a hurricane of steam and boiling hot rocks. The lateral blast flattened entire forests north of the volcano. At home in Oregon, 150 miles to the south, we heard the explosion like a series of sonic booms.

For several hours, pyroclastic flows of boiling hot rock, volcanic ash, and gases superheated to more than thirteen hundred degrees Fahrenheit poured from the crater. Melted snow and ice surged downstream, shoving rocks and logs in massive mudflows that filled the Toutle River valley. A column of ash rose twelve miles in the air and circled Earth for fifteen days. A thick dust of volcanic ash settled hundreds of miles downwind. I wrote the date with my finger through the ash on my car. Silica grit etched a permanent reminder onto the car's hood: May 18, 1980.

Before the 1980 eruption, the forests of Mount St. Helens were as green as any on the west side of the Cascade Mountains: Douglas-fir, western hemlock, mosses, lichens, and wildflowers. The eruption buried forests under hundreds of feet of shattered rock and ash. Mudflows choked streambeds for dozens of miles downstream. The blast zone was a moonscape. Ten days after the eruption, Fred Swanson entered the gray, smoldering landscape. "The place looked devastated," he later said. Muddy geysers erupted across a colorless blasted landscape.[11] While the mountain still grumbled underfoot and cement-heavy wet ash splattered their helicopter windshield, Swanson and three other scientists set foot on the newly born landscape. Fiery pyroclastic flows had created a sterile pumice plain across six square miles and two hundred feet deep. The cinder-covered land was still dangerously hot. In this choking-hot ash, something like a spiderweb caught Swanson's eye. It was the thin filament of a fungus evolved to respond to the heat of a forest fire. Heat from the eruption had triggered its spores to grow.

In the weeks following the explosion, scientists continued to visit Mount St. Helens, despite the danger of working in the blast zone of a smoldering volcano. They set up monitoring plots to mea-

sure changes in the landscape. One of the biggest changes they discovered was their own thinking. Expecting devastation, the scientists instead found evidence of survival. In protected spots at the
edge of the blast zone, they found broken but still living plants.
Runoff trickling across the volcanic ash had carved tiny channels down to buried but still living roots. A tiny green shoot nosed
through the ash—a lupine perhaps—with symbiotic bacteria to
draw nitrogen straight from the air. Life had survived.

It was a matter of luck for some, strategy for others. Life in
lakes and streambeds fared better than life on land. Ants lived; elk
died. Within a year, Pacific chorus frogs returned to wetlands; go-
phers mixed buried soil with surface ash; and a bouquet of fireweed and prairie lupine began to set seed. Even the relics of dead
trees helped rebuild streambanks and offered nutrients for coloniz-
ing plants and animals. Scattered remnants of the old landscape
would orchestrate the emergence of a new landscape. Jerry Franklin would call these remnants "biological legacies."[12]

Four years later, I flew over the volcano with my father in his
single-engine Cessna. Those hopeful biological legacies weren't obvious to me from the air. The north face still looked devastated,
fringed by the bleached bones of a million dead trees. Yet Spirit
Lake, which had received enough volcanic material to raise its elevation two hundred feet, shimmered blue beneath a thick layer
of floating logs. And the Toutle River had begun to etch a new
streambed through fourteen miles of debris-filled mudflow. A
struggling return of plant life was beginning to show green.

Now, several decades later, scientists continue to monitor the
return of life on Mount St. Helens, as they observe in real time how
ecosystems reassemble after a large disturbance. What follows a major disturbance is not a blank slate. "The role of chance and the influence of survivors can't be overstated," wrote Charlie Crisafulli,
a forest ecologist who spent his career studying Mount St. Helens.
"Had the eruption happened at another time of day or season, the
ecological consequences would have been markedly different."[13]

Mount St. Helens's renaissance has been uneven. Since 1980, subsequent eruptions have uncorked additional pyroclastic flows, and landslides have buried parts of the newly emerging forests. In the caldera, hot rock is still rising. Forty years after the eruption, mountain goats have returned, as well as bears and mountain lions. Elk have returned in such large numbers that much of the forest's regrowth is literally nipped in the bud. Change does not follow a predetermined path.

Soon after the 1980 eruption, much of the splintered forest of Mount St. Helens was logged for salvaged timber and replanted into fast-growing tree plantations. Revisiting the mountain de-

A seven-hundred-mile arc of volcanoes marks the eastern edge of the forest of fire and rain, including the Three Sisters at the boundary between Willamette and Deschutes national forests.

cades later, we see the plantation is approaching "merchantable" size, big enough to cut again. Next to the plantation, the unsalvaged landscape is a jumble of alders, willows, and wildflowers steadily overtaking volcanic pavements. Spiky firs encircle standing skeletons of their dead forebears. Life is abundant. This wild forest is the Mount St. Helens National Volcanic Monument—established in 1982 by President Ronald Reagan—a 110,000-acre laboratory to learn how life resurrects itself after a cataclysm.[14]

When scientists arrived on the volcano after the eruption, they did not find the complete devastation they had expected. The scientific ground had shifted under their feet. They were as surprised as the pocket gopher, shaken awake from her underground slumber, as she peered out toward a landscape of steaming rock and fumaroles. Now imagine the surprise of scientists, fresh pumice crunching under their boots, as they uncovered thin threads of burn-site fungi fingering through the ash. These scientists would follow such threads to a new way of understanding disturbance as a force of nature. Such cataclysmic disruptions are rare in the timescale of humans, and so we humans rarely have the chance to witness Earth's long-term response. Until now.

Evidence of volcanic eruptions is written into the geology of the Douglas-fir region—in vast lava fields, columnar basalt walls, and colorful petrified wood. Mount St. Helens is the most active of the Cascade volcanoes; she has erupted like this, and rebuilt herself, many times over millennia. We now understand that with every incarnation, the mountain is rearranged with different winners and losers, and life returns in different places and ways. Soon enough, a whole neighborhood settles in, eating, pooping, growing, and reproducing. Disturbance continues to reset the clock. Floods, drought, landslides, and fires create new neighborhoods with new winners and losers. The lesson of biological legacies is that whatever survives the cataclysm, no matter how small, is the beginning of a new start. It matters what you leave behind.

4

You Follow Steps toward a Very Long Life

Along the slopes of the Cascades, where the woods are less dense, there are miles of rhododendron, making glorious outbursts of purple bloom.

—JOHN MUIR, nineteenth-century naturalist and writer

Let's talk about succession. It's a touchy subject with certain human dynasties, and no more straightforward with forests. In a mythical world without disturbance, with no warring princes or forest wildfires, succession might be a predictable sequence of changes that gradually replaces one species, or heir apparent, with another waiting in the shadows. Here in the rainforest, we might see pioneering alders replaced by sun-loving Douglas-fir, which in turn would be joined by shade-loving western hemlock. Succession would depend on what species (or princes) were waiting in the wings, and on what mutually supportive relationships the trees (or the heirs) could rely on to support their claim to dominance. Like an ambitious and influential family, there's increasing evidence of network connections among trees that nurture growth and resilience in the younger members in the network. More often than not, however, the stately, predictable direction of succession is derailed by whatever disturbance is coming down the pike.

Only in classic ecological theory of succession does a forest march from seedling to old growth in a parade of sequential mile-

---

Sixty years after the original forest was clear-cut, this plantation in the Siuslaw National Forest has not regained its former stature.

stones toward an ultimate climax. Real life is messier, with lots of interruptions. Seral stages, as the phases of forest development are called, skip along at different rates. Fire, flood, or volcanic eruptions shuffle the deck of forest development, leaving whatever survives as a biological legacy for new growth.

I came of age as an ecologist at the University of Virginia under the mentorship of Bill Odum. His father, Eugene, wrote our textbook, the definitive *Fundamentals of Ecology.* The doctrine of the time, largely based on studies of abandoned farm fields and coastal barrier islands, was that succession was orderly, directional, and stabilizing.[1] However, naturally regenerated forests of the Pacific Northwest are not barrier islands or farm fields. Here, disturbance is always reordering and changing directions, and the ecosystem is almost never scraped clean without any legacy to build on. After all, a devastating volcanic eruption did not set the Mount St. Helens ecosystem back to the Big Bang. Some organisms manage to survive, one way or another. Disturbance happens unevenly, and it creates a jigsaw-puzzle landscape where seral stages intersperse and each stage of development shows up somewhere, at some time. The notions of succession, climax, and the balance of nature no longer apply; ecosystem development is a staggered process with no single, predetermined destination. Let's follow that process in our wild temperate rainforest.

We'll start from the moment after the forest has been struck by a stand-replacing wildfire. Don't look away in anguish. This moment, when the landscape looks blackened and broken, is not lifeless. This is what most of our old-growth forests looked like at the beginning of their lives. Depending on what buried seeds or roots managed to survive the fire, there soon will be fireweed, snowbrush, and rhododendron resprouting amid the charred remains of trees. These survivors, along with the legacy of large wood on the

ground, will slow
the pace of ero-
sion and enrich
the soil. Within
a few years, the
burned forest is
a sunny, shrubby
landscape beneath
the white bones
of standing dead
trees; the air is
alive with insects,
butterflies, and
songbirds. From
the forest's point
of view, this is its
early childhood,
the only time in
the long life of a
forest when it is
not dominated
by trees.

Something
in the burned for-
est signals wood-
boring bee-
tles to swarm in.
They come to lay

their eggs under the bark of dead and dying trees, where their larvae grow fat on the phloem. Woodpeckers, too, sense opportu-

A stand of fire-killed trees is an important, but rarely appreciated, stage in forest
development.

nity as they find dead trees for drilling nest holes and beetle larvae
for food. It's a short-term residency, however. As the beetles move
on to other newly burned sites, the woodpeckers follow, leaving a
neighborhood of nest holes ready-made for incoming songbirds,
owls, and kestrels.

Let's resist the urge to remove the burned timber and replant
the land with tree seedlings. Such salvage logging robs the forest of
the foundation it needs to reestablish. Logging the burned forest
destroys fallen wood essential for nutrient cycling, removes snags
that contribute to wildlife habitat, compresses soil, and opens the
ground to increased erosion. It kills tree seedlings and replaces
nitrogen-rich shrubs with tree plantations.

Left on its own, the blackened forest begins a long pageant of

----

Rain on snow can trigger high flows in rainforest rivers, such as this spring flood in
the Santiam River, Willamette National Forest.

forest development. Before young conifers take hold, sun-loving, nitrogen-fixing alders, mountain lilacs, and lupines pull nitrogen from the air to enrich the soil, where fire has volatilized much of the existing soil nitrogen. This early-seral-stage forest has among the highest diversity of plant and animal species of any point in its development, different in detail from the diversity in an old-growth forest, but just as important, as the forest ecosystem begins to rebuild itself from the ground up. Like all childhoods, this stage is brief and should not be rushed. It lasts only a few decades, until a young forest grows tall enough to close its canopy and shade its understory.

The young forest might remind you of other adolescents. Trees at this stage are growing fast, packing on biomass, and reaching for the sky. They compete for dominance. Their bark is pimply with resin blisters. Douglas-firs establish quickly in sunny, exposed soil. They grow tall and gangly, and they tend to elbow out the smaller trees and shrubs they overshadow. As tree crowns knit into a closed canopy, the forest floor goes from full sun to deep shade. Foliage is concentrated high in the canopy, so most sunlight goes to the tallest trees. In this race to the top, trees that fall behind drop to the floor, and their thin bodies decay quickly. It's brutal, like middle school. All those sun-loving shrubs, insects, birds, and amphibians fade away, and species diversity falls to its lowest level in the forest's life.[2] Young trees are vulnerable, since they have not yet developed fireproof bark or a pharmacy of protective agents. And because their roots aren't fully developed, young trees have only a tentative grip on the soil, unless they are anchored in a legacy foundation of heavy downed wood.

After one hundred years of growth, the mature forest no longer looks like a bunch of telephone poles. Competition has opened up crowded stands, and the remaining trees have lost most of their youthful gawkiness. They retain the uniformly pointed tops of their youth, and below the canopy tall, branchless trunks soar

like columns. After two hundred years, the mature forest takes on a more eclectic appearance: trees are thicker of girth, rougher of bark, and begin to sport lichens in abundance. At this point, Douglas-firs may have reached only two-thirds of their eventual height. They will continue to grow in height, girth, and crown for hundreds of years more, packing on wood and storing carbon.

As weaker trees die out, gaps open for shade-tolerant hemlocks, Pacific dogwood, vine maple, and other understory plants. The land has transformed from a fire-scarred landscape of snags to a grown-up forest rapidly building complexity in structure and function. Competition is no longer the main agent in tree death; now it is pathogens or environmental disturbance that topple weaker trees. The forest floor is beginning to pile up with larger logs with more decay-resistant heartwood that will remain in place for up to a century.

The death of some trees is a key feature in the maturity of the

---

Western hemlocks stand like a regiment arched over a long-remembered nurse log, bridging the past and future aboveground and belowground.

Douglas-fir forest. Ignoring this natural stage of development, forest managers sometimes use tree mortality to justify cutting trees from older wild stands in the name of forest health. "From an ecological perspective," Jerry Franklin told me, "such thinning is not only not needed, it will, in fact, interfere with the natural development processes that are underway. It will prevent formation of critical wildlife habitat and rebuilding stocks of coarse woody debris; it will negatively impact carbon stocks and slow the development of the understory and intermediate canopy of shade-tolerant species."[3] It is a different situation in the drier, fire-adapted pine forests east and south of the Cascades, where forest managers remove some younger trees from crowded, fire-suppressed stands to reduce fuel for wildfire, limit competition for water, and enhance biodiversity. Again, in the Pacific Northwest, the difference between west and east is like night and day.

Back in the Douglas-fir region, the old forest shows wear and tear from several hundred years of inescapable disturbance. Centuries of wind, fire, pests, and pathogens create distinctive individuals: their canopies sag, their tops are blown out, they have cavities, parasites, and heart rot. Aberrant branches show up like nose hairs. Yet these old forests continue to store massive amounts of carbon. After about four hundred years, old trees develop new, lower branches with additional foliage that fuels massive wood production. This is the stage where big trees become giant trees, and the snags and logs they produce when they die can last centuries. At around six hundred years, old trees begin to fall, creating gaps in the canopy where sunlight floods in. After centuries of patient waiting, shade-loving western hemlock, western redcedar, and Pacific silver fir make headway toward the sky. This is the classic old-growth forest, with the greatest structural diversity of any stage: giant Douglas-firs, western hemlocks in all sizes, standing snags, large downed wood, and a deep, continuous canopy. At eight hundred years, the old forest loses its oldest members as the last of the

original stand dies.[4] Even in death, these big, old trees contribute
to life in the wild forest. This is not the end, as we will see.

Let's return to the wild forest at its beginning, this time accompa-
nied by John Muir. While rambling through the forests of Oregon,
the nineteenth-century naturalist was dazzled by the wildflowers
that filled openings in the forest. He wrote of this trip in 1888:
"Along the slopes of the Cascades, where the woods are less dense,
there are miles of rhododendron . . . while all about the streams
and the lakes and the beaver meadows and the margins of the deep
woods there is a mag-
nificent tangle of
gaultheria and huckle-
berry bushes with
their myriads of pink
bells, reinforced with
hazel, cornel, rubus
of many species, wild
plum, cherry, and crab
apple."[5] Muir under-
stood that the shrub-
filled landscape that
so often follows a fire
was part of the glo-
rious fabric of the
forest.

   You can imag-
ine a similar delight
might be felt by mi-

A Douglas-fir cone tucks its seeds under what look like the feet and tails of mice
scurrying to hide.

gratory songbirds as they arrive in this magnificent tangle, after flying north for a few thousand miles from the tropical rainforests of Central and South America, to breed in open spaces of this temperate rainforest. Arriving here exhausted and depleted, birds such as olive-sided flycatchers and hermit warblers find a variety of flowering shrubs for foraging and nearby standing forest for hiding and nesting. If we are too quick to remove the burned forest and plant new trees, these migratory birds have no place to come home to.

That is because a tree plantation is decidedly not an early-stage forest. It is a monoculture of single-age, single-species trees with little to offer songbirds, woodpeckers, or other creatures of the rainforest. Douglas-fir trees planted at high densities tend to have narrow trunks, small limbs, and low resistance to wind, disease, and insects. Decades of managing for rapid reestablishment of conifers on cutover land has removed most of the flowering plants in plantations and therefore removed the insects that feed the birds that buzzed around Muir's delighted head.

Removing burned timber disregards how nature works. These days, most wild, early-stage forests are on public land, where wildfires have been allowed to reset the forest-development clock without extensive salvage logging and replanting. The result someday might be a forest with patches at all stages of development: tangled early-seral thickets, young beanpole groves, mature stands developing complexity, and old forests of elegant grace. Such a naturally generated, wild forest is a landscape molded by disturbance. It is complicated, improvisational, resilient, and diverse.

That was not how the forests appeared where I lived in the Coast Range during the timber boom of the 1970s and 1980s. Unlike the near-continuous block of publicly owned federal forest land that swaddles the Cascades from northern California to the Canada border, much of the Coast Range is privately owned. This land had

once held the biggest trees on the lowest land that were the easiest
to reach. After the timber was clear-cut, some of the low-lying land
was sold off to small rural communities. A few investors kept ownership and control of their timberland by building company towns
on their land. Valsetz, a company-owned timber town, was close
to our house as the crow flies, but two hours away by gravel roads.
Valsetz had a store, a post office, schools, and a formidable basketball team. As the old-growth timber was cut and hauled away,
many of the residents moved away, too. In 1984, the timber company razed the entire town and hauled away the remains of houses,
schools, roads, and the basketball court.

Our neighborhood wasn't owned by any big mill. Instead,
we had so-called gyppo mills where independent loggers milled
small material on contract. The mountains surrounding our small
valley were owned by various timber companies that cut and replanted trees at regular intervals like a crop. Behind our house and
behind the single flute note of a varied thrush, you could hear the
distant triple beep of logging machines. If the operation was closer
to home, you could hear chainsaws. Closer still, you could hear a
splintery wooden crack, a slow crescendo of breaking branches,
and a final thudding *whooomph* felt deep in your chest with each
falling tree.

Loggers in steel-studded boots negotiated the slopes behind
our house as easily as deer. They set choker cables around each log
and yarded it up to a landing. The logs etched scars in the moun-
tainside, lines that converged at the top of the slope. Loaded-up log
trucks roared out of the woods, heading to the mill. These eighty-
thousand-pound log trucks buffeted my Volkswagen Beetle along
the two-lane route to town. Sometimes, rarely even then, we saw a
truck pulling one single log, a massive cylinder six feet or more in
diameter—one single tree that needed several log trucks for transport to the mill. The biggest trees, too gnarly for lumber, ended
up in the pulp mill, coming out the other end as newsprint, card-

board boxes, or toilet paper. Occasionally, part of a giant log was fashioned into a welcome sign at the entrance to a logging town, erected with considerable civic pride. (Over the course of many decades, I've noticed these signs are periodically replaced by progressively smaller welcome logs or plywood signs.) Chip trains, heading to the pulp mill with tons of fresh wood chips, passed by our house each morning, smelling like pencil sharpeners. In those days, the mill was running three shifts a day. The acrid stench of cooking wood pulp blew into our valley ahead of each rain. Old-timers said it smelled like money.

Chip trains, clear-cuts, and log trucks were regular features of western Oregon and Washington in the 1970s and 1980s. Another was napalm, the explosive gel used in the Vietnam War to burn the jungle. Here in the Coast Range, napalm was tossed in pint-size grenades from helicopters to ignite piles of unwanted remnants of the harvested forest. Slash piles burned for days and lit up the hills at night with spots of flickering gold. Combat continued against the forest's wild childhood, killing pioneer shrubs with Agent Orange, a cancer-causing defoliant recommissioned from the Vietnam War. We were at war with the forest.

Agent Orange was a combination of the active ingredients 2,4-D and 2,4,5-T, with traces of a highly toxic by-product, the dioxin TCDD. Herbicides were an integral part of plantation forestry in western Oregon and Washington at this time. Long-lasting in the environment and linked to cancer, the herbicides were broadcast in a fine mist above the treetops by plane or helicopter, as they had been in Vietnam. And, similarly, they made their way into the water and bodies of people living near the forest.

During the 1970s, miscarriages reportedly spiked to three times the national average in small towns in the Oregon Coast Range, clustered around the time and place of summer spraying. Many of our distraught neighbors banded together as Citizens Against Toxic Sprays (CATS) and filed for a temporary restraining order to stop the aerial spraying of toxic herbicides. Regulatory agencies dithered, and spraying continued as laboratory studies piled up that confirmed herbicides' links to fetal death and kidney failure. A forestry professor from Oregon State University came to one of our meetings to calm our fears by filling a juice glass with liquid herbicides and drinking it down, presumably proving there was no risk because he didn't die on the spot.

Historian William Robbins recounts these years of agency denial and obfuscation in his aptly titled book, *Landscapes of Conflict: The Oregon Story, 1940–2000*.[6] Public pressure had forced the U.S. Defense Department to stop using Agent Orange in Vietnam in 1971; the U.S. Forest Service continued herbicide spraying in U.S. forests until 1983. Eventually, the courts banned the use of 2,4,5-T but allowed continued use of 2,4-D for aerial spraying on private timberland. In 2017, Lincoln County (Oregon) residents voted to ban this aerial spraying to protect the health and safety of the people and ecosystems in the Coast Range. A circuit-court judge ruled their ban was unconstitutional; the appellate court agreed. And so 2,4-D continues to be sprayed from the air on private industrial land, along with atrazine, triclopyr, and glyphosate.[7]

As the timber supply dwindled from private land, the Northwest forest economy became increasingly dependent on timber from federal land. National forest managers rose to the challenge to "get out the cut" at ever-higher harvest levels. At the same time, new environmental rules required them to draft comprehensive land-management plans that called for maximum sustained yield of timber as well as clean water, wildlife habitat, and opportunities for recreation. As harvests ramped up, the Willamette National Forest was the top timber-producing national forest in the United States, which left a lot of clear-cut ground that needed to be replanted.

Reforesting the Cascades and Coast Range was not the work of Johnny Appleseed. Replanting clear-cuts meant lugging heavy bundles of Douglas-fir seedlings down steep, muddy slopes in the middle of winter to plant trees across the face of a stump-pocked mountain. It was a hard job that attracted troops of college graduates who had fled the East Coast for ecotopia in the Pacific Northwest. Having named themselves after a tree-planting tool, the Hoedads began as a back-to-the-land cooperative of self-proclaimed "Freaks" centered around Eugene, Oregon. By the 1980s, they had grown into one of the nation's largest worker-owned cooperatives. Their superpower was a labor organization mixed with social justice and counterculture exuberance. It resulted in reforestation contracts that amounted to several million dollars over the decade. They planted billions of trees.[8] As individual Hoedads moved on to other careers, migrant laborers have taken their place replanting bare slopes on industrial tree farms.

Tree plantations, while not a good substitute for naturally established wild forests, were nevertheless an engineering marvel. With the land cleared of most other plants and any animals that might harm or compete with the trees, the idea was to direct maximum amounts of water, nutrients, and sunlight toward a predictable, efficient timber harvest. The seedlings themselves were from genetically improved stock of Douglas-fir, selected to grow fast and uniform. Planted all at the same time, the Douglas-fir trees would also be ready to harvest at the same time. The promise of a future with ever-increasing volumes of plantation timber made it easier to rationalize the immediate elimination of wild stands of old growth. The intent was to improve upon nature and create superior timber-producing national forests that would provide good-paying jobs for rural communities in perpetuity.

It worked for a while. For a few decades, timber towns benefitted from a flush of high-quality, old-growth sawlogs coming from the publicly owned forests. Timber was local, and mill owners were pillars in the community. Mill workers and loggers made good wages. Local mills sponsored school clubs, college scholarships, and softball teams.

However, prosperity could not be sustained, despite sky-rocketing timber harvests. Long before the spotted owl flew into the courtroom, lumber mills were laying off workers in favor of mechanization, as the supply of big trees to feed local mills kept dwindling. Unprocessed logs, fresh out of the Oregon and Washington woods, were being shipped across the Pacific to Japan and other Asian markets that could pay more for raw materials than local mills could afford. In 1988, Washington and Oregon exported 3.7 billion board feet of raw logs overseas, 24 percent of the year's total harvest, at a loss of fifteen thousand lumber mill jobs.[9]

Cutting down trees at a breakneck speed eventually led to landslides in the mountains, sediment in streams, and an ever-dwindling amount of old-growth habitat whittled into fragments, while herbicide sprays sickened rural communities. This was the battlefield on which the owl wars would be fought, pitting the ideals of ecotopia with the pioneering spirit of the Oregon logger. Our neighborhood in the heart of the Coast Range was a mix of New

---

Industrial logging has transformed parts of the wild forest into plantations of skinny young trees harvested every thirty to forty years.

Age idealism and traditional pioneer values, so you'd think we would be ripe for a showdown. But the owl wars were fought most ferociously in the federal forests of the western Cascades. There, the conflict led to burning down ranger stations, tree-sitting protests, and countless arrests. Here in the Coast Range, state and private timberland kept lumber mills running business-as-usual, with relatively few restrictions.

But even in the Coast Range, changes were occurring that would transform forests and forest communities. New machinery, such as the one-man feller-buncher, made it possible to cut and stack four thousand trees in a day with the flick of a joystick.[10] Automated mill equipment took over the job of sawing logs into rough lumber and trimming rough lumber into finished boards. Local kids and their parents were losing their jobs, and the Coast Range forests were being rapidly cut down. Succession was stuck in reverse; people and forests were getting poorer.

About one-third of the forestland in Oregon and Washington is privately owned, and most is located on highly productive land. Since the 1980s, investor-driven corporations have bought large tracts of private forest with the expectation of high returns on investments, similar to other corporate stocks. Waiting for a clear-cut to grow back as a mature forest is not in the business plan of these private industrial landowners; trees grow slower than money. As a result, millions of acres of industrial timberland are kept in permanently arrested development, harvested and replanted every thirty or forty years on land that is cleared of everything except stumps and saplings. This land has the capacity to grow a huge, wild forest, but it will take a change of ownership and at least one human lifetime for the land to recover from decisions made in corporate boardrooms far from the temperate rainforest.[11]

5

You Glimpse What
It Means to Be Old

Humans necessarily bring an anthropocentric perspective to the work of defining old growth. But we are creatures of imagination. With empathy and respect, we may ask: how would an old-growth forest define itself?

—KATHLEEN DEAN MOORE, environmental philosopher and writer

For much of the last century, the old trees of the temperate rain-forest have been the subject of changing attitudes, from a limitless resource to an irreplaceable treasure. How did this change occur in human reckoning? To make sense of this shift, it's helpful to look at the environmental crises making headlines in the 1960s: oil spills in California, a polluted river on fire in Ohio, DDT in human breast milk, and poisonous smog killing people in American cities. The prosperity that showered over the United States following World War II had come at a high environmental cost.

Rachel Carson warned us about a poisonous silent spring (1962); Paul and Anne Ehrlich warned us about an exploding population bomb (1968); Garrett Hardin warned us against the tragedy of the commons and our virulent selfishness (1968).[1] There were tragedies of Jim Crow laws that ostracized Black Americans; flagrant violations of Native American rights; assassinations of national leaders; and a horrific war in southeast Asia. All this while the world's superpowers were threatening each other with nuclear weapons. Despair and anger heated the globe and burned the cit-

---

A new promise is whispered to an old Douglas-fir.

ies. From my childhood home across the Potomac River from Washington, D.C., I watched smoke rising from fires burning the capital. I was fifteen.

Then, on Christmas Eve 1968, something changed. The Apollo 8 astronauts sent the world a photograph of Earth rising above the surface of the moon. We saw our shared and only home, alone in space. While I was learning how to drive, I was also learning how to stand, with millions of others, to care for this planet. We pledged our new allegiance on April 22, 1970, the first Earth Day. I sewed a green-striped, theta-emblazoned Earth Day flag that we hoisted up the flagpole in front of

A pileated woodpecker chisels holes that will become homes for the next wave of residents in the old forest.

T. C. Williams High School, three miles from the Pentagon, and replaced the stars and stripes. We marched across the river to the National Mall in Washington, D.C., and joined the National Environmental Teach-In. There was a young forest growing in me, elbowing out competing ideas and reaching for the sky.

Across the Potomac, the power and influence of environmentalism were having a moment. Months after taking office in 1969, Richard Nixon visited the devastation of the Santa Barbara oil spill. He had been elected president just as the public began to embrace environmental concerns. At that time, politicians in both parties were falling over themselves to claim the mantle of environmental advocacy.[2] Republican and Democratic lawmakers worked together to plan and promote that first Earth Day, in which twenty million Americans took part, almost 10 percent of the population.

I remember watching the TV news that night. Reporters characterized Earth Day mostly as a bunch of hippies enjoying a sunny day. Walter Cronkite described the event, saying, "Its demonstrators were predominantly young, predominantly white, . . . [and] curiously carefree." It's true, it was not a diverse crowd on the streets that day. But Earth Day was not just a day in the park. As carefree as it seemed to the buttoned-up TV newsmen, the environmental movement would not fade away. The crisis was undeniable, as it is now. We weren't carefree; we believed we could change the world, and we were ready to try. One week later, Nixon invaded Cambodia. A few days after that, members of the Ohio National Guard killed four students at Kent State. A few weeks after that, Mississippi Highway Patrol officers killed two students and wounded twelve more at Jackson State College.

Tangled up with protests against the war in southeast Asia and for civil rights at home, the environmental movement continued to grow from teach-ins to political change, and people felt empowered to influence government policies. In a brief moment of Congressional bipartisan action, groundbreaking legislation

emerged—the Environmental Protection Agency (1970), the Clean
Air Act (1970), the Clean Water Act (1972), and the Endangered
Species Act (1973). Integral to these new laws was the National Environmental Policy Act (1970), with its mandate to report the environmental impacts of all large, federally funded projects and to
allow the public to review plans and sue for compliance to these
new laws.

Before the ink dried on these laws, they were challenged. Following World War II and decades of mostly unregulated logging
of wild forests on private land, the Pacific Northwest timber industry had turned to the publicly owned national forests to fuel
the nation's rapacious postwar building boom. Responding to pressure from politicians and industry lobbyists, the Forest Service increased its harvest levels to the full extent promised by intensive,
high-yield forest management. (The Forest Service itself doesn't
cut commercial timber. It identifies areas of public forest to be
cut, then offers those so-called timber sales for private companies
to bid on.) Congressional delegates from Oregon and Washington amended appropriation bills to increase federal timber sales in
the Pacific Northwest.[3] These actions pushed against the intent of
the new environmental protection laws and fueled conflict in the
Douglas-fir region that continues to this day, with old forests at the
crossroads.

Unlike hurried youthful forests that can crowd into bare landscapes in a matter of decades, old forests seem timeless, immutable.
In older stands, many Douglas-firs are more than five hundred
years old; cedars and yews might be closer to one thousand. These
trees know a thing or two about surviving. Each tree has established itself under conditions in the distant past that favored its
success, and each has lived through innumerable changes in those
conditions ever since.

Mortality is key to the for-
est's life. An initial legacy of left-
over snags and fallen trees sheltered
the forest's early-seral childhood.
The crisscrossed poles of dead saplings nourished the young-forest
stage. The wood that piled up in
the mature forest began to build
a foundation of heavier logs and
snags with a significant amount of
decay-resistant heartwood. Slowly
decaying logs store tons of water
like giant sponges and provide rich
planting beds for new trees. They
anchor the ecosystem, feed and water it, and build a foundation that

will carry survivors forward after the next big disturbance. Mortality ensures resilience.[4] However, mortality can take a long time.

Walking into an old Douglas-fir forest, you are struck by the
personality of each old tree. Battered by time and weather, they
have multiple arms, bushy sideburns, and snaggle-toothed tops.
Deep roots and thick bark help protect old trees from drought,
fire, and insect attack. Epicormic branches, sprouting from the
dormant buds of long-gone limbs, develop into massive second-
ary trunks with short clusters of unruly branches. Branches criss-
cross from forest floor to treetops. More than one hundred species
of moss and lichen thrive in the branching arms of these old trees.
Even young hemlocks can root themselves a hundred feet or more
above the forest floor, along with an assortment of birds, bats, in-

Weighing less than three pennies, a hermit warbler flies twenty-five hundred miles
(one way) each spring from the mountains of Central America to breed in the old
forests of the Pacific Northwest.

sects, fungi, and mammals, many of which could live nowhere other than in the embrace of these old trees in these old forests. Indeed, there are entire ecosystems up there.

Down here, it's shady and cool. Little birds—Wilson's warblers, hermit warblers, varied thrushes—flit through the understory hunting bugs. Many migratory songbird species are declining in numbers, so this sight of warblers and thrushes during breeding season is a hopeful sign. Landscape ecologist Matt Betts and colleagues have found that some songbirds find refuge against a warming climate in the cool old forest. These are tiny birds with tiny bodies that can be sensitive to local differences in temperature or humidity. "Old forests buffer those differences," Betts said. "Their deep, layered canopies stay cooler in summer compared to younger forests or plantations." Also, the wide variety of plants in an older forest extends the time when flowers and insects emerge, so when migratory birds arrive from the tropics to breed, they have a better chance of finding food.[5]

All these features—deep canopies, complex architecture, moss-laden logs, and snags—create a richly productive ecosystem with an abundance of varied habitats for plants and animals dependent on the old forest. Some are as small as the red tree vole, which lives its entire life in the high branches of older Douglas-fir trees. Up there, the tree vole feeds mostly on fir needles, despite the nasty-tasting chemicals that repel most other herbivores. Tree voles peel away the unpalatable resin ducts and eat the inner needle like a banana. If you find a wad of small threads that looks like a tangle of hair, you know that red tree voles are busy overhead.

Unlike the hard-to-see tree vole, the Douglas squirrel is hard to miss. This tireless, chatty seed harvester gets an entire chapter in John Muir's 1894 book, *The Mountains of California*. Another namesake of David Douglas, the squirrel is as widespread in the Cascades and Coast Range as in the mountains of California. Muir describes the Douglas squirrel as a "fiery, sputtering little bolt

of life. . . . He is the mocking-bird of squirrels, pouring forth mixed chatter and song like a perennial fountain. . . . Probably over half of all the ripe cones of the spruces, firs, and pines are cut off and handled by this busy harvester."[6]

Another vocalist in the forest, the pileated woodpecker, announces itself with a laughing holler like Woody Woodpecker's and a matching red cap. An important architect of forest habitat, this big woodpecker chisels holes in snags with head-banging efficiency, holes that eventually become homes for owls, bats, and flying squirrels. In search of insects in decaying wood, pileated woodpeckers help break apart dead wood for the next round of decomposers.

The red tree vole is well adapted to live its entire life in the high branches of old Douglas-fir trees, but is unable to survive without an old forest. (Drawing based on photo by Michael Durham, https://www.flickr.com/photos/oregonwild)

It's always startling to see Roosevelt elk in the forest. They are huge, the largest elk in North America, with antlers up to four feet from head to tip. Herds of them seem to appear out of nowhere, often at forest edges, where they browse and battle. During mating season, you can hear combatting males clatter their enormous headgear, a sound like rattling dishes. And they bugle—not like a trumpet, but with a growl that ends in a scream you can hear a mile away.

One more creature of the old forest you are not likely to encounter: the elusive, probably mythical Sasquatch. Based on Native American lore, Bigfoot has become a playful, hairy symbol of this temperate rainforest. Two laws on

---

Because it doesn't hibernate, a Douglas squirrel hoards a big supply of seed cones in underground caches, where some forgotten seeds will eventually sprout into new trees.

the books in Washington legally protect Sasquatch on the grounds
that, if it does exist, it's obviously endangered. Seems logical.

Seriously now, let's consider carbon sequestration, an important
feature of the old forest and just as invisible as Sasquatch. Carbon
dioxide is pulled from the atmosphere by an abundance of forest
plants and stored (sequestered) in trees, in the woodpeckers' snags,
in logs scattered across the forest floor, and in moist, rich forest
soil. The Pacific temperate rainforest contains more biomass than
any other region on Earth, which means it has an abundance of
containers in which to store carbon. And because the trees in our
rainforest grow so big and tall, these long-lived trees continue to
absorb additional carbon as they continue to grow for six hundred
years or more.[7]

In 2021, multibillionaire Elon Musk announced he would do-
nate $100 million toward a prize for the best technology to remove
carbon dioxide directly from the atmosphere and store it in a safe,

---

Roosevelt elk prefer a mix of old-growth forest and open meadow where they can
find both shelter and food.

cost-effective way. Could we submit our rainforest? The so-called XPRIZE goes to a technology that will remove one thousand metric tons of carbon dioxide a year, can be scaled up to remove one million metric tons a year, and will have the capacity eventually to remove one gigaton of carbon dioxide per year and store that carbon for at least one hundred years. No problem! The rainforest in Oregon's Coast Range and western Cascades alone removes thirty million metric tons of carbon dioxide a year, stores it for several hundred years, and has been doing so for millennia in shady sanctuaries of peace and solitude, for free. Consider that the world's forests already remove two gigatons of carbon each year. Perhaps we should stop cutting them down and collect the prize money.[8]

The Save the Redwoods League, established in California in 1918, motivated "people of good will" to purchase stands of old-growth redwoods in order to save them from being logged. Their efforts managed to preserve dozens of old redwood groves in northern California that were eventually consolidated, first as state parks and then, in 1968, as Redwood National Park. The park is a testament to both avarice and generosity; the trees are magnificent, yet the park protects less than 5 percent of the original redwoods. Unlike the redwoods of California, the big Douglas-firs of Oregon and Washington have never had a league of defenders to protect the most magnificent groves. In the fifty years between 1918 and 1968, while the redwood park was assembling, most of the oldest, tallest coastal Douglas-firs were cut down.

However, big, old trees still exist in the Douglas-fir region beyond the reach of timber sales. The rainy west side of Olympic National Park and surrounding wilderness areas protect several world-class trees within a stone's throw of each other, including world-record contenders for the largest western redcedar, largest Alaska yellow-cedar, and largest Sitka spruce, along with runners-up for the world's largest western hemlock and some humungous Douglas-firs and big-leaf maples. These world records are constantly challenged by modern tree-climbing explorers, who have helped discover and protect remnant stands of old-growth temperate rainforest in their quest for the biggest, tallest, and oldest trees in the world.

How do you measure a tree that is thirty stories high and growing in a layered canopy obscuring most of its height? Forest scientist Steve Sillett began his exploration of tall trees by climbing to the top and dangling a tape measure down to the ground. Sillett's early searches for the tallest coast redwoods were chronicled in Richard Preston's 2008 book, *The Wild Trees.* Now Sillett and collaborator Michael Taylor use prisms and narrow-beam LiDAR (Light Detection and Ranging) to locate some of the tallest trees in the world. LiDAR uses lasers to create three-dimensional maps of Earth's surface to reveal the jagged tops of forests.

Branching out from their famous explorations of redwoods, Sillett and Taylor began searching for the world's tallest remaining Douglas-firs. LiDAR and prisms can't magically transport you into remote, rugged coastal valleys. So, the explorers bushwhacked through tangled understories of crisscrossed logs and flesh-ripping devil's club into the steep canyons of western Oregon and Washington. Their efforts were rewarded. In 2021, Taylor located and documented several previously unrecorded big, old Douglas-firs, including the tallest in the world: a 326-foot Douglas-fir standing in a steep coastal canyon in southwestern Oregon.[9]

Before plundering logging began in the early twentieth century, Douglas-fir may well have been the tallest tree species on the planet. The world's tallest tree known today is a 381-foot coast redwood in California that Taylor codiscovered. The tallest Douglas-fir ever reliably measured was even taller: 393 feet tall before it was toppled in the 1930s. A ten-foot-diameter cross-section of this fallen giant is on display at the Wind River Arboretum. It has 865 rings.

Steve Sillett is the Kenneth L. Fisher Chair of Redwood Forest Ecology at Humboldt State University. His pioneering studies of canopy ecosystems are far more valuable to science than to Guinness World Records. After years of climbing with ropes and tools to examine every inch (millimeter, actually) of life in the vast ecosystems of ancient trees, Sillett has developed a way to assess—from ground level—the rate of biomass accumulation, and therefore carbon storage, held in each giant tree, in addition to its height and the diversity of life in its branches. And by every measure, the trees of the Pacific temperate rainforest are exceptional. But what he *doesn't* see worries him, particularly in the Douglas-fir forests. "I don't see mature trees coming up to replace the oldest trees," Sillett told me. "We are losing the elders to drought, fire, and beetles. And we've cut so much, there's so little left, and nothing to replace them. We have to protect our potential elder trees if we hope to have forests in the future."[10]

Great height is one distinction; great age is another. Only two dozen tree species in the world are known to live beyond one thousand years without human support. Eight members of this exclusive millennial club are native to western North America: bristlecone pine, giant sequoia, coast redwood, Douglas-fir, western redcedar, yellow-cedar, western juniper, and western hemlock. For adventurers and scientists searching for the oldest trees on Earth, there's a database, called the Old List, of the world's oldest trees whose ages have been officially measured. (The oldest tree on the list, an unnamed bristlecone pine, is confirmed to be 5,077 years old.)[11] Such ancient trees have inspired some of the oldest human stories. The Buddha attained enlightenment beneath the Bodhi tree. Stories from the Salish Sea tell how the madrone's tangled roots keep the earth from falling apart. Yggdrasil, the Tree of Life at the center of Norse mythology, also attempts to hold the world together during the earth-shattering war of Ragnarǫk.

These cautionary tales come to mind in the modern search for the world's oldest trees. One way to measure a tree's age is by counting the rings in its trunk: one ring per year of growth. Tree rings also show trends in temperature, precipitation, drought, and volcanic activity used to reconstruct past climates. The procedure is a bit like a biopsy, except that it uses a drill that removes a pencil-thin core from across the grain of the tree trunk. In 1964, a researcher curious about the age of a gnarled bristlecone pine in eastern California got his tree-corer stuck in the tree. To recover the lost corer, a helpful park ranger cut down the tree. When the researcher began counting rings on the stump, he realized he had just cut down the oldest living tree ever recorded at that time; it was almost five thousand years old. The stump has been named Prometheus.[12]

Far beyond championship age and stature, old trees encourage long-term thinking about relationships, consequences, and the effects of change. Consider the well-documented relationship among truffles, squirrels, and owls in the old-growth Douglas-fir forest. Truffles are a type of mushroom, the reproductive parts of certain fungi associated with the roots of many kinds of forest trees, including Douglas-fir. Unlike most other mushrooms, truffles hold their spores in fleshy bundles underground; they depend on animals to disperse those spores. Squirrels are among the forest's most avid mushroom-eaters, and northern flying squirrels, in particular, seem to eat nothing as much as truffles. Drawn by the musty truf-

fle aroma, a flying squirrel parachutes down to the forest floor at night, unearths the truffles, and later expels spore-filled pellets throughout the forest. These fecal pellets are not just packages of poop; they mix fungal spores with nitrogen-fixing bacteria and microorganisms necessary for trees to access nutrients. Flying squirrels, in turn, supply most of the diet of northern spotted owls.

So, the owl depends on the squirrel that depends on the fungi that support the forest that supports them all.[13] It's a tidy food web woven from intricate relationships in the old-growth forest.

Adapted to life in the complex architecture of old forests, northern spotted owls fly silently through the maze of canopy branches. These owls don't migrate; instead, they move up and down in the deep forest canopy to adjust to temperature changes through the seasons. Their food and shelter are closely tied to mature and old forests. When wildlife biologist Eric Forsman chose to study the northern spotted owl as an undergraduate at Oregon State University in 1969, the bird was a bit of a mystery. There were few recorded sightings, and no description of nesting or feeding habitat. The northern spotted owl seemed elusive, unless you knew exactly where to look. Forsman learned where to look and how to call the owls to him using their punctuated four-beat call: "Who—

The northern spotted owl hunts the flying squirrel that hunts the truffle that holds the spores of fungi that knit together the forest that shelters them all.

hoo-hoo—hoooo." Possibly because the owls had encountered
so few humans in these old forests, they were unafraid of people.
"You can walk right up to them," Forsman recounted in his oral
history. "The media loved them because you could take field trips
out and these owls would come flying in and literally land on your
camera."[14] The little owl with the liquid-onyx eyes became a media
darling.

Forsman followed the obliging owls to their nests, built
among the layered branches of old trees deep in the old forest. He
watched throughout long nights as the owls delivered food to their
chicks—mostly flying squirrels, tree voles, and woodrats, creatures
that also depend on old-forest habitats. Forsman calculated that
spotted owls hunt across a vast home range, up to forty-five hundred acres of mature and old forests. And these old forests were
scheduled to be logged. "We had just started finding these birds,
and every time we'd find another pair, we'd find out that there
were three [timber] sales planned in that area," he said. "Then
you'd start to look around, and there would be blue paint everywhere."[15] Blue paint marks the boundary of a timber sale.

Within a few years, the Endangered Species Act had become
law, soon followed by other legislation aimed at protecting species
and habitats. By the 1990s, the fate of the northern spotted owl,
and the old forests in which it lives, became a national conservation cause fought in the courts and logging towns. The laws pre-
vailed, and in 1994, most old-growth logging was halted on federal
land within the range of the northern spotted owl, with the intent
of halting the decline in owl numbers.

It worked for a little while, until a new challenge appeared
in the forest—the barred owl. Native to the eastern United States,
barred owls have moved north and west across the continent, easily
adapting to local food and shelter options. Upon arrival in the Pacific Northwest, they settled comfortably in the old forests of the

northern spotted owls, using the same nesting sites and hunting the same flying squirrels, voles, and wood rats, and a lot more besides. Barred owls are adaptable; northern spotted owls much less so. If the old forest becomes too limited, barred owls can switch to suburban woodlands or to younger second-growth forests adjacent to the northern spotted owls' dwindling old growth. Barred owls eat whatever they can catch—frogs, skunks, fish, or other birds. And they are fiercely territorial, driving spotted owls from their nesting territories and preying on their chicks. A barred owl's home range is less than one-tenth the size of a spotted owl's home range, which means that one spotted owl pair might have to run the gauntlet of ten pairs of barred owls.[16]

The barred owl has become a new top predator that most old-forest animals have not evolved to evade or compete with. Its increased presence could affect more than the survival of the northern spotted owl.[17] As an experiment, wildlife managers have tried killing hundreds of barred owls (also a legally protected species) to save the northern spotted owl. It was a wrenching experiment for many biologists, and it brought up difficult questions about the consequences of human attempts to control the natural world. However, killing barred owls has slowed the decline of northern spotted owl populations in those habitats where the two owls overlap. It's worth remembering it was the loss of old-forest habitat that pushed spotted owls to the brink of extinction. The barred owl, elbowing its way into the last vestige of old-growth forest, has placed an even higher premium on what remains of large expanses of wild conifer forests here in the range of the northern spotted owl.

Northern spotted owls no longer respond to researchers' hooting calls. Perhaps the owls have learned to stay quiet and invisible; or perhaps they are gone. In any case, the owl changed the fate of this rainforest, and now the fate of the owl is at risk. The story of a forest is a story of constant change delivered by unpredictable disturbance. Nothing stays the same for long. An interloping owl, a changing climate, or a shift in public values can change everything. Forests are not machines that can be programmed to suit our needs. They are complex living systems where everything gets complicated by everything else.

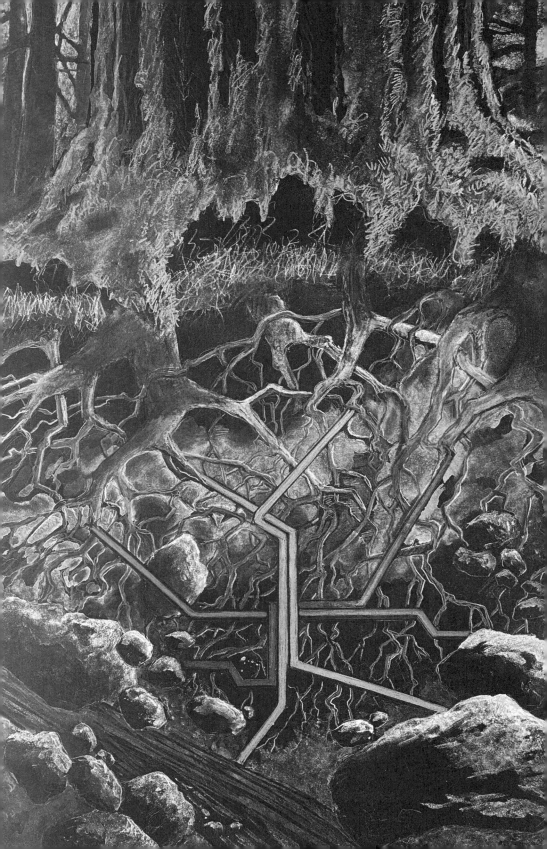

6

YOU EXPLORE THE
FOREST UNDERWORLD

You can't have a forest this magnificent without taking the time.

—MARK HARMON, forest ecologist

What happens when a tree falls in the forest? The immediate answer is that a sudden change in air pressure creates a wave of energy that can be perceived by tiny structures in the human ear. But that's not the point of this old riddle. It really asks: how do we know something exists if we can't observe it? Before we fall too far down the rabbit hole of quantum mechanics, let's consider other things that happen when a tree falls in the forest.

Much of the answer is beneath your feet. The forest's recycling center is down here, along with much of its energy exchange and community diversity. Structurally, the underground is as complex as the aboveground, just harder to see. Bacteria, fungi, algae, and tiny invertebrates busily break down and rebuild the essential elements of life. Transactions of goods and services are carried out in a world like Alice's Wonderland, with wild characters in a bustling underground metropolis. These inhabitants live fast, with high turnover, and they require lots of energy from aboveground. More than half a tree's energy from photosynthesis goes to feed this huge microbial network that keeps the whole forest functioning.[1] When a tree falls in the forest, its legacy echoes for centu-

---

Most of the forest's energy goes to running this busy underground transit system.

ries as the process of decomposition recycles nutrients for the next stage of life. Let's take a look.

Weighted down by heavy snow or pushed by a gust of wind, the bulk of an ancient Douglas-fir has become more than its roots can hold. Slowly at first, the tree leans toward an unbearable tipping point. As gravity takes hold, the tree's roots rip from stony soil. Its top crushes against neighboring trees. With the sound of gunfire, branches break as the giant body crashes through layers of canopy, taking smaller trees down with it. The tree shatters across the forest floor. The earth shudders. A rain of needles and shattered lichens shower down. A skylight opens. This fallen tree is our entry into an unseen world of life after death in the forest. When a tree falls in the forest, it begins a slow progression of decay played over centuries, much of it too obscure for a human ear or eye to perceive.

The afterlife of trees is an idea that until recently was invisible to most people, like the unheard sound of a tree falling when no one is there. Early logging of the region's wild forests left lots of stumps and broken wood scattered across hillsides. Only the best logs were hauled out, leaving unwanted trees in heaps of slash and residue. Such waste was thought to be a fire hazard and an obstacle to planting the next crop of trees, so thrifty forest managers called for burning piles of dead wood to clear the land for replanting. They also required loggers to pull woody debris out of streams to clear the way for fish migration. The forest of the future was to be swept clean and tidy after each harvest.

However, back in the 1970s, a handful of forest ecologists who had been tripping over tangles of dead wood in the old forest paused to realize that all this stuff they were stumbling over was an undeniable part of the forest. Working at the at H. J. Andrews Experimental Forest, they calculated that fallen trees occupy nearly 30 percent of the old-forest floor, and that these downed logs are doing more than simply blocking paths: fallen logs hold large amounts of water and nutrients such as nitrogen, calcium, and phosphorus. As the logs decay, these nutrients become concen-

trated, increasing more than threefold during the process of decay. It is the forest of the living dead. This realization turned their forest research upside down.

A live tree, despite its heavy bulk, has a slender claim to being alive: a thin layer of cells beneath the bark generates tissues to transport water and nutrients to growing roots and leaves. The living tree's massive trunk and branches are mostly carbon, and mostly dead. A really dead tree, on the other hand, squirms with life: insects, fungi, bacteria, and protozoa that in turn make way for slugs, salamanders, and sapling trees. The tiniest creatures of the living-dead forest—the decomposers—recycle old life back to the basics for new life: water, carbon dioxide, and nutrients essential to the health of the forest and to the planet. Dead trees take a long time to recycle, and new life shows up in a long, slow procession. To understand exactly how long, forest ecologist Mark Harmon began to monitor the decomposition of trees in a research study designed to take two hundred years.[2]

The forest at ground level is a lively place, including carabid beetles, yellow-spotted millipedes, and chanterelle mushrooms.

In the mid-1980s, Harmon's research team dragged hundreds of logs of different sizes onto forested sites at the Andrews Forest, in a bold example of very long-term ecological research. And then they watched. They watched as beetles chewed intricate galleries into the dead logs. They watched as hitchhiking spores on those beetles spread eager fungal filaments throughout the galleries. They watched as pioneering microbes took up residence in the wood and began to rearrange the furniture.

Some animals eat wood straight up. Others carry a cocktail of decomposers in their guts to digest chewed wood and share nutrients with their hosts. Carpenter ants and woodpeckers don't eat wood at all, but the holes they excavate offer a place for wood-eating fungi to enter the tree. A clattering of dampwood termites descend on the log, digging new channels into the wood. A lively banquet of decomposers nibbles away until the log crumbles into geometric blocks of cubical brown rot or into damp, spongy fibers of white rot.

What looks like rot to us is alchemy to the forest. When a tree falls in the forest, it delivers tons of carbon and other nutrients locked away in durable packages of cellulose and lignin. Both brown-rot and white-rot fungi can digest the log's tough cellulose (the material that makes plants fibrous), and white rot also digests *really* tough lignin (the material that makes wood rigid). In the process, the fungi release carbon and nutrients essential for continuing life in the forest. Soil bacteria are equally important, decomposing fallen branches, leaves, and dead fungal filaments relatively rapidly. Together, these salvage operators reassemble the pieces of old life into new life.

Of course, all this takes a very long time. At the decomposition study site at the Andrews Forest, scattered logs lie at rest, each with its own metal tag: the name, rank, and serial number of those who died for science. Some of the logs have plastic tubs and little tents protruding from their bodies, sheltering instruments that measure the slow exhalation of carbon dioxide. As in everything about this forest, the process varies depending on the species of

tree, the size of the tree, the decomposers that move in, and where the tree falls in the forest. After decades lying flat on the forest floor, the silver fir has slumped into a soft, mossy mound. The western redcedar lying next to it shows signs of decaying sapwood, but its heartwood backbone is still firm against the ground; it could take more than two hundred years to soften this rot-resistant log. Trees that fall into streams decompose even more slowly, in part because submerged and waterlogged wood lacks the oxygen needed by many decomposers. Unearthing the importance of dead trees to living forests is what Harmon calls "morticulture."[3]

When it comes to life in the forest soil, mushrooms get all the attention. They are the tasty, attractive fruiting bodies of fungi, while the rest of the organism sprawls unnoticed belowground. Neither plant nor animal, fungi reside in their own kingdom. A fungal kingdom suggests metaphors of culture and fantasy, where magicians turn rocks into rubble and grapes into wine. This is a kingdom with in-

---

When a tree falls in the forest, a procession of life-forms recycles the old parts of the tree into useful components for new life.

dustry, alliances, and relationships played out at the very foundation of life. Fungi are death-eaters and agents of reincarnation. They
make our bread, beer, and many medicines. They clean up oil spills;
they decompose dead bodies and turn them into soil.

The cell walls of fungi are not made of cellulose, as are the
cell walls of all plants; they are made of chitin, the same crunchy
stuff found in insect shells and fish scales. But fungi aren't crunchy.
Most forest fungi take the shape of hollow, feathery filaments
called hyphae that thread together in a gossamer, netlike mycelium
hidden in soil or rotting wood. Mushrooms and truffles are mere
temporary structures designed to scatter spores for reproduction.
The storybook toadstool in the woods might be the only visible
sign that a large network of threadlike mycelium sprawl underfoot.

Fungi may have hooked up with algae over five hundred million years ago in order to leave the water and colonize land. When
that happened, everything changed. These pioneers filled the atmosphere with oxygen, they scavenged nutrients from rocks, and
they enticed herbivores to follow them onto the terrestrial world.
Today, 90 percent of land plants still depend on associations with
fungi to provide most of the nitrogen and phosphorus that plants
require. Those associations are made through filaments known as
mycorrhizae that connect plant roots to the underground network
of fungal mycelium. Those fungal threads that fingered through
the ash at Mount St. Helens were busy knitting together nitrogen-
fixing bacteria, plant roots, nutrients, and water to build a new,
complex network called soil.

Biomass (the total of all living matter) found underground
amounts to about 20 percent of all the biomass in the forest. However, underground life uses more than half of all the energy produced aboveground by photosynthesis. "There is a tendency to
think that soils support the trees," forest ecologist Jerry Franklin told me. "In fact, soils require large inputs of high-quality energy that they get from green plants." If you remove the green
trees, he says, the soil loses energy; fungi and other parts of the un-

derground network wither; and the soil begins to die. "The soil has got to have green plants growing in it to live!" Franklin says. "When herbicide-soaked plantations burn in a wildfire, there is no resilience in the form of green plants that can immediately sprout back and lifeboat the soil."[4]

Underground life is promiscuous. Mycorrhizal fungi have relationships with many species of plant, and plants have relationships with many kinds of fungi. Multiple partners ensure survival. Douglas-fir trees can associate with as many as two thousand species of fungus throughout their life spans, and those fungi make up a significant part of forest soil. If you sink your fingers into the cool, damp duff of an old Douglas-fir forest, you're likely to unearth something that looks like an old, tattered shawl. It is a mat of mycorrhizal *Piloderma,* an abundant fungus that, in an association with bacteria, knits a web of mycorrhiza that boosts carbon and nitrogen uptake for an entire community of trees, shrubs, and ferns.

Mycorrhizal fungi are the resource-exchange ministers of the underground. Plants attract mycorrhizae using chemical signals that steer fungi toward a plant's roots and signal fungal spores to germinate there. In turn, the mycorrhiza signals the plant to suppress its immune system so the fungus can enter the plant root and grow inside. Fungal threads snake through pores in the soil where bulkier tree roots can't go, in search of nutrients they absorb with enzymes that trees don't possess. Mycorrhizae grow faster than tree roots and can extend one hundred times farther. As they finger through soil, mycorrhizae collect nitrogen, phosphorous, and water, and they share those resources with a cooperative of trees in exchange for the carbon-rich sugar the trees produce during photosynthesis. Through this symbiosis of plants and fungi exchanging food and water, forests develop a circulatory system, a valuable service overlooked by most mushroom-hunters, who seek only fungi's showy reproductive system via chanterelles, boletes, and shiitakes.

There's treasure everywhere on the forest floor. Consider this small stump where you are sitting. It looks like an upholstered hassock, with smooth bark covering its top, inviting you to sit and rest. Root-grafted stumps like this indicate where a cut tree is kept alive by neighboring trees. Instead of dying, the stump's raw cut heals over with bark-like tissue that protects the stump from decay and keeps the roots alive to continue transporting nutrients from the soil to surrounding living trees. This little hassock was made by the underground economy beneath your feet, by water and nutrients traveling uptown into trees and carbon traveling downtown into soil.

Suzanne Simard, a professor of forest ecology at the University of British Columbia, and her colleagues illustrated such an underground network by mapping connections among fungi and tree roots in a stand of Douglas-fir in British Columbia. Their map looks like the route a bicycle messenger might take on a busy day in Seattle. It shows mycelial connections across a bustling underground marketplace, linking trees and transferring nutrients, even between trees of different species. Using tracer elements, the map shows almost all of the trees in this stand are linked together. Big, old trees have big root systems, more root tips, and more carbon flowing into the mycorrhizal networks, so the biggest, oldest trees in this stand are linked to the greatest number of other trees. Simard refers to these matriarchs as the "Mother Trees."[5]

Simard has been rooting under forests for most of her career. It began with an early observation she made while working in the Douglas-fir forests of British Columbia. She noticed that naturally seeded trees in undisturbed forests were thriving, while the seedlings growing in adjacent plantations were stunted and weak. Twentieth-century forestry dogma held that trees competed for water and nutrients, so foresters would eliminate all competition to ensure fast growth and high yield. Simard's observations, and years of meticulous research, challenged that assumption.

Simard's research, published in *Nature* in 1997, was one of the first to show how intertwined fungi and roots transfer resources and chemical signals. Plants emit hormones and defense signals. Other plants detect these signals and rally their own defenses. In revealing an underground network where trees share resources and information, Simard and her colleagues have shown that forests aren't just collections of individuals competing with one another, but a network of symbiotic relationships. Biologist Merlin Sheldrake describes it as mutually assured survival.[6]

While mycorrhizal fungi manage the transfer of nutrients, saprophytic fungi do most of the decomposition. Together with bacteria, these rotters digest cellulose and lignin, the bricks-and-mortar of living plants, to release nutrients that would otherwise be locked up and unavailable to plants. Without this recycling, Earth would be smothered in dead bodies and shit. And indeed, during the Carboniferous period, 308 million years ago, when lignin and cellulose were novel enough that few organisms had evolved a way to digest them, forests piled up without decomposing. Eventually, they became fossil fuel.

A more nefarious fungal group are the parasitic fungi that attack living trees. One badass parasitic fungus is the sweet-sounding honey mushroom, *Armillaria,* implicated in many conifer deaths in the Douglas-fir region. Rather than grow as thin threads of mycelia, *Armillaria* also grow thick, black cords called rhizomorphs, whose gnarled fingers can extend for miles through the soil in search of roots to kill and wood to eat. Eastern Oregon is home to the world's biggest tree-killing *Armillaria.* Its filaments spread underground for more than three square miles. DNA tests confirm this is one single fungus and perhaps the largest single living organism in the world. If you could gather it all into a pile, it would weigh upward of thirty thousand tons, and it has been growing for thousands of years.

Here's another superlative: the temperate rainforest holds the most diverse collection of moss species anywhere in the world. Mosses are easily overlooked as a wall-to-wall carpet that cushions boulders and logs, so you're going to have to get on your knees to peer into this miniature forest. Mosses are key to nutrient cycling in the rainforest; they can also hold ten times their weight in water, which makes them an excellent holding tank for drip irrigation in the forest. Without moisture, mosses dry up in a crisp, crinkly mat with no signs of life, no metabolism, no respiration. They shrivel up and wait, sometimes for decades. Life-giving moisture resuscitates them, and they unfurl like tea leaves in a cup of hot water. Mosses survive in dim forest light by squeezing extra chloroplasts into their cells to catch every ray of sunshine that filters through the canopy. Their demand for nutrients is low, so they can grow on almost anything—rocks, trees, rooftops—and feed on dust and rain.

Moss leaves are a mere single cell thick, with cell walls that contain a pectin-like substance to thicken rainwater and hold it in the moss's spongy structure. Hikers know that a moss-covered log can stay soggy for weeks after a rain, dampening the surrounding rainforest and the seat of your pants. Invertebrates depend on the life-giving moisture suspended in mosses, as do fungi. A blanket of moss above increases the density of mycorrhizae below. Remove the moss, and the density of mycorrhizal root tips diminishes. Once stripped from a tree, moss takes decades to grow back. This worries bryologists, who see the multimillion-dollar harvest of forest moss—sometimes legal, often not—outstripping the ability of moss to reestablish.

It was moss growing in downtown Portland, Oregon, that helped finger the source of toxins in the city's air back in 2015. At the time, the city was using three air-quality monitors to measure daily air pollution. When the instruments detected cadmium and arsenic spiking to six times the safe levels, the three scattered monitors could not identify the source. But U.S. Forest Service research ecologist Sarah Jovan had a plan. Because moss absorbs what it needs from air, and because Portland is covered with moss, Jovan used moss as a dragnet to examine air quality in every corner of the city. After analyzing 346 tufts of the ubiquitous green Lyell's bristle moss plucked from Portland's trees, Jovan's team drew a moss-plus-cadmium map of the city. It pinpointed two toxic hot spots with a level of precision that was impossible for the air-quality monitors. A similar study in Seattle used moss to locate unusually high concentrations of toxic metals—arsenic, cadmium, chromium, cobalt, nickel, and lead—in a neighborhood of predominantly low-income people of color, a low-cost way to begin to correct seemingly invisible environmental injustices.[7]

No one knows moss better than Robin Wall Kimmerer, a botanist, writer, and member of the Citizen Potawatomi Nation. Kneeling down to enter the miniature moss world, Kimmerer finds inspiration for the larger world. "We're busy looking for biological, ecological and cultural solutions to climate chaos," Kimmerer told the *Guardian* in 2021, "but mosses, which have been with us ever since they arose, 400 million years ago, have endured every climate change that has ever happened."[8] In her thoughtful meditation, *Gathering Moss: A Natural and Cultural History of Mosses,* Kimmerer thanks the mosses for helping other plants and animals to flourish. "The patterns of reciprocity by which mosses bind together a forest community offers us a vision of what could be," she writes. "They take only the little that they need and give back in abundance."[9] We humans might practice such gratitude and reciprocity.

Peer into any mossy world with a hand lens and you'll see a carnival of tiny animals. In a mossy sample plucked from the rainforest floor, you see carabid beetles with fierce-looking pincers patrol the top of the moss canopy; rotifers wheel around like egg beaters in tiny suspended pools; and water bears float through the

foliage like balloons in a Macy's parade. Let's pause here to look at water bears, little manatees floating in a raindrop sea.

Tardigrades, as water bears are technically called, are a group of organisms all their own, a branch split from the family tree five hundred million years ago. They each look like an animated croissant with button eyes and eight chubby legs with tiny claws to grip onto branches of moss. The size of a grain of sand, tardigrades live almost anywhere on Earth—the top of mountains, the bottom of hot springs, the Antarctic, and the temperate rainforest. And they have superpowers, which scientists have been pushing to their limits. If their mossy ocean dries up, tardigrades can replace the water in their cells with a protective sugar and shrink into a dried-up ball. In this state of suspended animation, tardigrades can survive being boiled, encased in a vacuum, and frozen to almost absolute zero, for years at a time. They have survived exposure to outer space and pressures on Earth up to six thousand times the pull of gravity. And with a drop of water, they have returned to their living, floating, tubby little selves.

The size of a grain of sand, a water bear lives suspended in the moisture of forest moss and in many more unlikely places.

Other invertebrates, including nematodes and rotifers, whose entire world exists in a few inches of forest floor, use suspended animation to survive disturbances they cannot escape, shutting down during drought, flood, or temperature extremes. Some soil organisms are more mobile. Springtails, for example, burrow deep into the moss, where they feast on fungi and occasionally spring about like little kangaroos. Oribatid mites, by contrast, are sluggish, appearing to be tiny drops of blood going nowhere. The Andrews Forest has perhaps the world's most extensively studied oribatid mites, where as many as 120,000 individuals, of more than sixty species, have been counted in a square meter of forest floor. It's amazing what you can find when you take the time to look.

Where life is abundant, life in the soil is abundant, too. As numerous as stars in the night sky, the number of organisms in a teaspoon of soil from a Douglas-fir forest might include one hundred million bacteria and miles of fungal hyphae, along with countless soil-dwelling species that spend their quiet lives chewing, shredding, and pooping to make more soil. Don't snicker: arthropod feces make up the lion's share of matter in the organic layer of soil. Fecal pellets are a highly concentrated mix of nutrients, stuck together with mucus from the guts of these tiny creatures. Practically every particle in the top several inches of forest soil has passed through somebody's gut. You can't buy this soil in a bag.

Soil arthropods are the Meals-on-Wheels of the underground, distributing nutrients through the soil to microbes that have limited means of getting around. Sow bugs (the only crustaceans adapted for land-only living) shred fallen needles and twigs, scattering crumbs to less mobile microorganisms. Legions of these smaller creatures are hunted by centipedes, pseudoscorpions, and carabid beetles. Larger arthropods are, in turn, hunted by salamanders, birds, and shrews. In this rainforest soil, everybody eats. Soil organisms enrich the soil that feeds the fungi that grow the trees that create the forest that feeds the soil. And so it goes.

You don't need a hand lens to find a banana slug. Up to eight

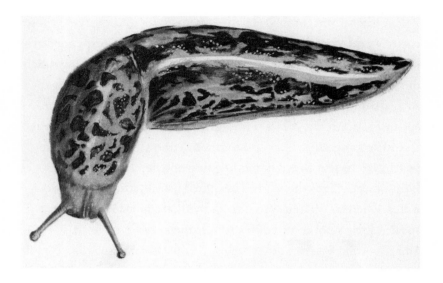

inches long and the color of overripe bananas, banana slugs are the second-largest slug species in the world. They are voracious recyclers in the forest, gobbling up dead plants, animal waste, and mushrooms that they turn into nutrient-rich fertilizer for the forest soil. Banana slugs have no interest in your garden strawberries. They are native inhabitants of the damp temperate rainforest, famous for gliding along a slime trail they create for themselves by mixing dry granules of mucus with the forest's ample moisture. These mucus granules can absorb up to one hundred times their volume in water, which explains why it's so hard to wash slug slime off your hands. Banana slugs are simultaneous hermaphrodites: they can mate as male or female. Their male equipment is prodigious, almost as long as the slug itself, among the biggest penis-to-body ratios in the animal kingdom. Another superlative for the temperate rainforest.

    A favorite animal for many at the H. J. Andrews Experimen-

Banana slugs, native inhabitants of the damp forest floor, depend on their slime to glide, climb, deter predators, and accumulate bits of food.

tal Forest is the rough-skinned newt. Seemingly unafraid, it sashays across the forest floor on baseball-gloved feet, flashing its orange belly to flaunt the potent toxin in its skin, a toxin many times deadlier than cyanide, the same chemical that is in pufferfish. The toxin is *in* the newt's skin, not *on* it, so it is possible to touch a rough-skinned newt, but it's not advisable. In a wonder of evolutionary adaptation, a few daring garter snakes have managed to eat rough-skinned newts, toxin and all, and live to tell it. And rough-skinned newts manage to eat banana slugs, mucus granules and all.

All this drama is acted out when a tree falls in the forest. Eventually the rotten log, soft from decomposition and rich with compost and water, becomes a seed bed for new trees, nurturing seedlings with reservoirs of time-released nutrients. The seeds of shade-loving western hemlock in particular have a hard time taking root anywhere except in the softening mulch of a decaying log. In an ancient cycle, seedlings sink their roots down into the past and send their shoots up to the future. Bottom to top, the forest is a city that never sleeps.

A rough-skinned newt is adorable to look at, but possesses a neurotoxin that can, if ingested, cause paralysis, asphyxiation, and death.

7

You Drift to the Top
of the Canopy

What has always amazed me about climbing into the canopy is that there are
such dramatic differences, both in microclimate and biology, just 30 meters
above the forest floor.

—NALINI NADKARNI, forest biologist and canopy explorer

Over the years, scientists have used all sorts of contraptions to
study the tops of trees—balloons, ropes, drones, and chainsaws. For sixteen years, arbornauts at the Wind River Experimental Forest used a construction crane to lift them into the canopy of an old Douglas-fir forest. I rode to the top a few times
on Wind River's canopy crane, hard-hatted and strapped in to
a bright-yellow gondola that dangled from the end of a 279-foot
boom, 285 feet above the ground. On this field trip, we will use our
imaginations.

Looking up, our route to the canopy disappears in a misty
crosshatch of limbs. Colors above us shift from green to blue as
short-wavelength light is scattered in the vapor of air. Our complex
vision may have evolved from something like the tardigrade's button eyes, a single light-sensitive cell, but clearly there was more we
wanted to see. Human eyes developed to spot predators and prey
hiding among a hundred thousand shades of forest green. And
even now, with eyes that can distinguish at least one million colors in a blink, humans see more shades of green than any other

---

The layered scaffold of a mature rainforest seems to reach to the moon.

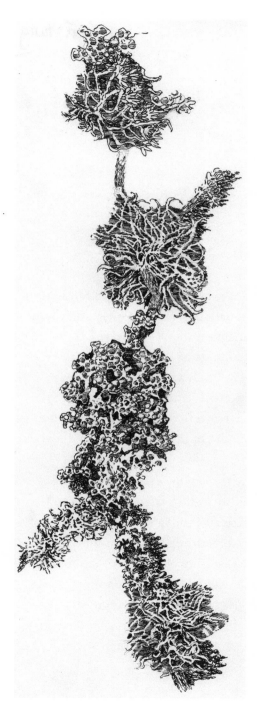

color. Poet John Keats complained that Isaac Newton had "destroyed the poetry of the rainbow by reducing it to a prism."[1] Really? It seems that understanding what it means to be a rainbow has enriched art *and* science ever since Newton.

Now you are riding a rainbow toward the top of the rainforest. You lift off from a shamrock carpet of bright-green oxalis. A sapphire-winged Steller's jay eyes you suspiciously; a golden-crowned sparrow flits by your ear; a red-breasted nuthatch tootles in the distance. You drift above red huckleberry, vine maple, and a row of young hemlocks lined up as straight as the nurse log that bore them. A Pacific wren dashes mouselike underneath you. Tree

---

Bits of various lichens decorate a stem of big-leaf maple, signaling good air quality.

trunks rise on either side like monumental columns. The bark of these old Douglas-firs is a rugged landscape in miniature, with flat bark-plateaus separated by deep bark canyons. Spider webs span the canyons; the bark plateaus are forested with moss.

Your eyes might distinguish more than a million different colors, but your nose can distinguish more than a *trillion* different scents. Here in the forest, you breathe air perfumed with volatile hydrocarbons known as terpenes, among them lemony limonene and turpentine-scented pinenes. Terpenes react with air to produce fragrant aerosols; each species of tree has its own chemical identity. The greater the diversity within the forest, the more varied the biochemistry within this tide of atmospheric fragrances. Douglas-fir has a noticeable fruity, resinous scent, while western redcedar smells like pineapple, and western hemlock like celery. Such chemicals help protect trees against invasions of bark beetles or other pathogens, but many have a sweeter effect on us.

Rain activates aromas. Consider the April showers scent created when rain hits dry ground. As air becomes humid, the bacteria *Streptomyces* produce a biochemical called geosmin on porous soil, rocks, and city pavements. Raindrops splatter geosmin and other volatile plant oils, creating the heady, mineral fragrance called petrichor. The human nose can detect tiny amounts of geosmin, and so too can springtails, tiny arthropods that squirrel around in forest moss and litter. As it turns out, *Streptomyces* bacteria need springtails to disperse their spores, so they use geosmin to lure the little arthropods for the job. Why humans evolved to swoon over petrichor is anybody's guess.

A forest is a good place to go whenever you want to breathe. Several studies show how terpenes and other tree aerosols can inhibit inflammation from human diseases such as bronchitis, chronic obstructive pulmonary disease, and osteoarthritis. Epithe-

lial cells that line the surfaces of your body are designed to absorb
these aerosols. Some researchers theorize that when we breathe in
fresh forest air, we breathe in airborne antibacterial and antifungal
chemicals that plants use to ward off disease, and our bodies respond by increasing our own natural immunities.[2]

The Japanese practice of *shinrin-yoku,* oddly translated as
"forest bathing," is touted to restore the physical and psychological
health of humans through a "five senses experience" of the forest
environment. It's more like breathing than bathing in the Western
sense. Now that you are in the forest, slow down, breathe deep, and
see if it works for you.

This is a quiet forest, not at all like the clattering, howling din of
tropical rainforests. Here in the temperate rainforest, you can pick
out individual voices in a soft chorus of sounds. A Pacific wren
squeaks like rubber on wet glass. An olive-sided flycatcher orders
what sounds like, *"Quick, three beers!"* A varied thrush responds
with an unhurried, single-note flute in a haunting minor key. Entrepreneurs market recordings of forest sounds to people with in-

somnia. Ecologists use sound recordings to monitor animal biodiversity. In the old days of field biology, we used nets and traps to
monitor ecosystem activity, or we recruited volunteers to count all
the creatures they could find within a certain time and place. Forest animals are hard to spot, evolved to be seen only when they
choose to be seen, so our old-school census-taking was laborious
and incomplete. Now, sensitive microphones and machine-learning
tools record evidence of forest activity all day and night, creating a
soundscape of birds singing, streams flowing, and earth rumbling.
Even underwater, recorders can detect the sound of aquatic insects
scraping algae and cobbles rolling in the current.

A new era of spotted owl research uses this sonic technology.
The decline of spotted owl populations has made it much harder to
locate the owls, and worse, when researchers have been successful
in calling in a spotted owl, barred owls would come, too, and acted
like schoolyard bullies toward the responding spotted owl. Damon
Lesmeister, a wildlife biologist and head of the Pacific Northwest
Bioacoustics Lab, developed a type of artificial intelligence to distinguish the sounds of barred and spotted owls without drawing
them into contact with each other. So successful was this AI re-
connaissance that Lesmeister's team has expanded its audio recognition to more than forty-five species of bird and mammal living in
the range of the spotted owl, providing a rich audio portrait of life
in the old forest.[3]

Water drips through the canopy in soft percussion, as raindrops
bounce down as throughfall or stream down the trunks as stem-
flow. Either way, as moisture percolates through layers of canopy,
it washes detritus from needles and bark. The protracted drip-drip
through the forest canopy is rich in accumulated nutrients, fun-

The long, single note of a varied thrush seems to haunt coastal forests.

gal spores, and nitrogen-rich detritus. Some of this biological rain nourishes fluffy feather boas of *Isothecium* mosses and neon-green tufts of wolf lichen (*Letharia vulpina*) that cling to the bark of old Douglas-firs. In the wet, western side of Olympic National Park, big-leaf maples wear moss like heavy fur coats.

Mosses and lichens are epiphytes ("epi" means "top"; "phyte" means "plant"). They are not parasites. They live off airborne water and nutrients and use the trees only for support. The largest old-growth trees support the most massive epiphyte communities. Over centuries of growth, these aerial communities will become a large proportion of the forest's total biomass. As they die, epiphytes decompose into rich pockets of soil as much as a foot deep that can be perched one hundred feet or more off the ground. Sometimes the host tree itself will produce thin roots from its branches to tap these rich canopy soils, roots that are indistinguishable in form and function from underground roots.

Epiphytes can help protect the forest against drought by soaking up moisture and storing it in their own tissues or in the canopy soils. The wet-weight of rain-soaked epiphytes in an old Douglas-fir forest has been estimated to be three to five tons per acre of forest.[4] As you drift up, notice the aerial garden of huckleberry bushes ripe with berries tucked in with licorice ferns and western hemlock seedlings, all rooted in soil high above the forest floor. Tiny clouded salamanders can live their entire lives in this airborne jungle; aquatic crustaceans swim in raindrop pools; a marbled murrelet cradles her single egg in a thick bed of moss. It is a forest within a forest.

As you lift up through layers of conifer branches, use your X-ray vision to peer into the needles that surround you. On the underside of a fir needle are microscopic openings called stomata. They look like tiny lips, puckered as they inhale carbon dioxide and exhale oxygen and water. On warm, dry days, as more water is exhaled through the stomata, the tree sucks more water up from the roots. You can hear water moving through the belly of a tree with as simple a tool as a stethoscope. Choose a thin-barked tree more than six inches in diameter. Move the stethoscope against the trunk until you hear a faint snap-crackle-pop as water mixes with air on its long ride up to the treetop. Lots of snapping means lots of air bubbles are interrupting the flow of water. As the day gets hotter, needles close their stomata to conserve water; photosynthesis stops. In the cool of the night, tree roots and associated mycorrhizal fungi begin to draw water from deep beneath the tree. Water again flows upward all night. By dawn, the tree is refreshed, the forest floor is damp, and the water level in a tiny nearby stream may have dropped measurably.

Now look deeper into the needle and you might see wispy filaments of fungi so small that they squeeze *inside* needles to work as guards for the tree against ever-changing threats of pathogens. These so-called endophytes ("endo" means "inside"; "phyte" means "plant") are known mostly from research on agricultural grasses where they produce toxic alkaloids to ward off insect attack. Few people thought to look elsewhere for these tiny, enigmatic organisms until mycologist George Carroll looked up from hunting mushrooms on the forest floor and wondered what might be happening high above his head. A lot was happening, as it turns out. Carroll and his tree-climbing team found endophytes to be ubiquitous residents inside conifer needles and speculated (accurately, as it turns out) that endophytes might be as common among plants as mycorrhizae.[5]

In exchange for sugars that the tree produces by photosynthesis, most endophytic fungi produce alkaloid compounds that can poison foliage-eating insects. In conifers, endophytes function like a SWAT team, spurred into action at the first nibble to toughen their chemical and physical defenses. Rapidly evolving, like the insects they protect against, endophytes can quickly develop new defenses in response to new tactics developed by foliage-eating insects. This is an ingenious advantage for very long-lived trees that can't run away or evolve defenses rapidly enough to protect themselves against swiftly adapting threats. Rapid-response endophytes do the job for them. A single tree can contain many different kinds of endophytes with a wide range of poisons and repellants, providing broad protection at very low cost to the tree.

An endophytic fungus lives its life crammed inside one single cell in the fir needle, doing very little for years. However, at the moment the fir needle is nibbled or dies (needles live for about seven years), the endophyte releases its spores to disperse and settle in fresh needles elsewhere in the canopy. There, new endophytes grow and wait, dormant and dangerous, killing time until an insect dares to nibble.

Among this cryptic group of fungi, one has made headlines. Endophytes in Pacific yew trees produce a chemical, paclitaxel, that can be deployed to ward off fungal diseases that infect cracks in the yew tree's bark, similar to the way your immune system works to protect against infection in a wound. As it turns out, paclitaxel does something similar in humans as it does in yew trees. Paclitaxel is the source of Taxol, a powerful anticancer drug that is particularly effective against rapidly dividing tumor cells in humans.[6]

Other tiny forest organisms provide other magical abilities, such as spinning life-giving nitrogen out of thin air. Although it makes up 78 percent of Earth's atmosphere, nitrogen gas (N_2) must be converted into ammonium (NH_4) before it can be of any use to plants. Nitrogen gas is held together by strong bonds that require a lot of energy to convert into usable molecules that feed plants. A bolt of lightning can deliver this much energy (and sometimes does), as can the industrial combustion of a lot of fossil fuel (in the manufacture of chemical fertilizers). But for much of the history of life on Earth, pulling nitrogen from the atmosphere has been primarily the job of bacteria and archaea (a distinct group of microorganism); the equally important job of returning nitrogen gas to the atmosphere has fallen to fungi.

Here in the rainforest, red alder and associated nitrogen-fixing bacteria convert nitrogen in the air into usable nitrogen that is stored in nodules on the alder's roots. This alder-bacteria duo can also extract calcium and phosphorus from rocks, releasing mineral nutrients to further feed tree roots. In addition, free-living cyanobacteria can fix nitrogen directly by their own photosynthesis. With factory-like efficiency, single-celled cyanobacteria photosynthesize during the day and fix nitrogen at night. After usable nitrogen is absorbed by plants, and those plants are eaten by animals, the nitrogen cycle is completed by fungi that decompose plant and animal tissue and release inorganic nitrogen back to the atmosphere. *Voilà!*

In the early-seral forest, nitrogen fixing is the work of fast-growing woody species such as red alder, lupines, and mountain lilac that bank enough nitrogen in the soil to jump-start a fast-growing young forest. Eventually, however, the mature forest outgrows these early-successional shrubs. It needs another nitrogen source. Fallen trees contain substantial amounts of nitrogen, especially heavy logs in mature forests. With help from decomposers, these logs provide a slow drip of nitrogen into the forest. Even more nitrogen enters the older forest from the top, by way of *Lobaria oregana,* a lichen that decorates high branches with its lacy ruffles. Cradling free-living cyanobacteria, bits of *Lobaria* rain down from the canopy and scatter the forest floor with what looks like leaves of savoy cabbage. Each crinkled bit of lichen delivers life-nourishing nitrogen captured from the atmosphere at the treetops. Intensive industrial forestry cuts off these natural sources of nitrogen at both ends of forest development: by clearing the land of old

logs, removing nitrogen-fixing shrubs, and harvesting young trees
before old-forest lichens can develop.

A lichen is an unlikely partnership between a fungus and a green
alga and often includes a cyanobacterium—a three-way partnership that involves members of three separate kingdoms. These partners don't fall easily into each other's arms; they seek a mate with
just the right chemistry. Once the right partner is located, fungal
threads surround the algal cells and enclose them within a mesh
(called a thallus), which is something that neither fungus nor alga
can create on their own. Cyanobacterium joins the partnership,
adding nitrogen fixing to the lichen's bag of tricks.

Now joined as a lichen, the so-called mycobiant (fungus) and
so-called photobiant (alga and cyanobacterium) create a dynamic
partnership with remarkable powers: the photobionts produce sugars from sunshine; the fungal mycobiont can crush rocks to release
minerals. Lichens are so successful they can be found throughout
the world, from cold, dry Antarctic valleys to searing-hot Chihua-
huan deserts. Similar to our hardy friends the water bears, lichens
can be completely dried up for decades and come back to life with
a few drops of water. They have even survived strapped to the outside of the orbiting space station.

How do three different species from three different kingdoms live in the same body *and* reproduce? Unlike plants that can
produce seeds to grow new plants, most lichens can't grow new lichens. Often it is the fungus that produces spores that produce another fungus that must go out and find another algal partner. Al-
ternatively, a resourceful lichen such as *Lobaria* produces tiny
granules of already allied fungi and algae. These granules split off
from the main lichen, and if they settle on welcoming surfaces,
they can grow into new *Lobaria* lichens without the uncertainty of
locating a new compatible partner.

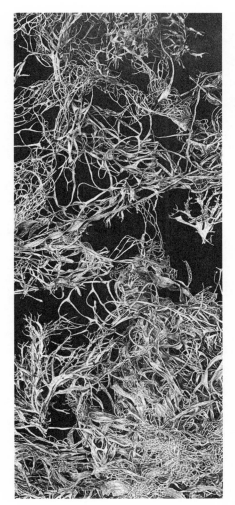

Lichens are excellent indicators of air pollution. Like mosses, lichens absorb what they need from the air, and much of what they don't need, too, including heavy metals. Most lichens are particularly sensitive to air pollution. Red alders that grow where air is unpolluted shimmer with silver-white lichens plastered like thick paint on their bark. Alders growing in polluted areas have much less lichen; their undecorated bark is plain tan. Where the air is clean, a variety of lichens add a paint box of colors to the rainforest. Look around: cyanobacteria create blues and purples; algae bring yellows and greens; fungi provide white and silver.

Like many unorthodox ideas in science, the idea that lichens were a partnership of fungi and algae was anathema to the scientific lions of the nineteenth century. However, the idea had an early and completely overlooked champion in Beatrix Potter. Long before the children's book author followed Peter Rabbit down a rabbit hole, Potter studied mushrooms and lichens. With an eye

---

Stretchy when wet, like an elastic band, *Usnea* lichen drapes in long, tangled nets from forest branches.

for detail and pattern, she observed these organisms under her microscope, germinated fungal spores, and meticulously drew what she saw. But because Potter was a woman (and an artist to boot), she was shut out of scientific societies, and her attempts to publish her findings were blocked. What remains of her mycology work are hundreds of paintings of mushrooms and lichens so detailed and accurate that they are still used 140 years later to illustrate particular species and processes.[7]

The forest floor is now far below as you continue to drift upward past an odd-looking branch arching from the trunk of a massive old Douglas-fir. This epicormic branch has sprouted from a dormant bud beneath the bark and now reaches up like a second trunk. Some epicormic branches in turn sprout more epicormic branches, in a carousel pattern characteristic of old Douglas-fir canopies. A deep, complex canopy develops multilevel platforms where small mammals, birds, and insects find homes. Red tree voles can spend their entire lives up here, feeding on little else than fir needles.

The mature forest canopy is like an apartment building with a different ambient temperature and humidity at each level. Mosses are most common on the damp, dark lower floors. *Lobaria* prefers the lower-mid-canopy. Hair lichens, such as *Usnea,* are more common in the sunnier, wind-tussled upper crowns. Birds and insects follow suit, vertically distributing themselves within these distinctive high-rise apartments.[8] You drift pass a snarled mass of dwarf mistletoe that encircles the trunk of a big western hemlock. Peek inside and you might see a flying squirrel nestled asleep. Then . . . *pop!* A mistletoe seed suddenly shoots out, propelled by explosive water pressure that builds inside the berry as it ripens. The sticky seed clings to any surface it hits, including this canopy explorer.[9]

Floating up through the forest, you breathe in oxygen released by the green plants below. In this old forest, like most in this region, Douglas-firs make up the topmost canopy where shade-loving western hemlock and silver firs fill lower gaps. The upper canopy holds most of the needles, absorbs most of the solar energy, and accounts for most of the carbon and water exchanges between forest and atmosphere. Carbon is a gregarious atom, able to form bonds with other atoms in dazzling varieties of organic molecules, often in long, chain-like molecules that contain a lot of energy. When these long carbon chains break, a burst of energy is released, the rocket fuel for life on Earth.

Carbon circulates in the atmosphere and in rocks, oceans, green plants, and us. It is ubiquitous across the cosmos, one of the first elements that formed after the birth of the universe. You are made of carbon. So are stars, plants, animals, and fossil fuels. Most of the carbon on Earth is stored in rocks, and it moves in a complicated shell game called the carbon cycle, sometimes as slow as a mountain, sometimes as fast as a puff of smoke. In an earlier chapter, we followed the long carbon cycle as part of the planet's geologic groundwork. Now let's dip into the short carbon cycle and its mysterious handmaiden, photosynthesis.

The short carbon cycle happens within life rather than within rock. The short version is: plants eat sunshine and burp out oxygen. The mystery lies in those long carbon chains. Almost all the oxygen we breathe is made by green plants as they rearrange molecules of water and carbon dioxide to form energy-storing molecules of sugars and life-giving oxygen. Note the color green. It is chlorophyll, a green pigment packed in the cells of green plants and in the membranes of certain bacteria. Sunlight zaps these emerald-colored packets (called chloroplasts) and excites the proteins within. Electrons fly in all directions, splitting the H from the two Os in water. Oxygen is released into the atmosphere. Hydrogen enters a weird stage of molecular sleight-of-hand, as it loops through long chains with rearranged carbon; each compound fuels the formation of the next. In green plants (and in algae and cyanobacteria, too), the sequence of events ends with long carbohydrate chains of sugars that serve as batteries storing long-term energy for building a forest. Out of thin air comes life on Earth, running on solar power.

Photosynthesis stores energy by remixing atmospheric carbon dioxide into carbohydrates; respiration releases that stored energy through metabolism in daily rhythms of air flow and sap flow. Globally, this fast carbon cycle is so closely tied to green plants that you can map a growing season by the way carbon dioxide fluctuates in the atmosphere. In spring, when plants are growing and absorbing carbon dioxide, the amount of atmospheric carbon dioxide dips; in winter, when plant growth slows and annual plants decay, atmospheric carbon dioxide concentrations increase. Earth is breathing.

Similar to bacteria in lichens, these emerald-green chloroplasts are descendants of bacteria that gave up their independence more than a billion years ago to relocate inside plant cells. Now consider this: magnesium makes chlorophyll green; iron makes hemoglobin blood red; otherwise, the molecular structures of chlorophyll and hemoglobin are almost identical. Both are made up of four elements—carbon, hydrogen, oxygen, and nitrogen—arranged in a circle around a hub. Chlorophyll, with its magnesium hub, captures light. Hemoglobin, with its iron hub, transports oxygen. Think about it: you might be just an atom away from photosynthesizing.

Suddenly, you pop out at the very top of the forest. It is rough terrain, dry, spiky, and wind-swept, with a big, bright sky all around. Up here, the forest gets the full extent of sun and wind. At night, it gets cold. It looks like sunlit tundra, and nothing at all like the deeply shaded layers of lower canopy. Vaux's swifts circle above,

having spent the night huddled together inside a big,
hollow, chimney-like snag.
The top of the forest undu-
lates in peaks and hollows,
poked here and there by jagged broken treetops. Each
tree species has a distinctive crown: Douglas-fir has
a chandelier of large epicormic branches, western redcedar has a veil of drooping
branches, western hemlock
has a comb-over top.

The forest works like an
air-conditioner, pulling water from the soil and releasing it as water vapor through
millions of needles. At the
top of the canopy, another
sort of air-conditioning helps
cool the planet, where treetops form a rough, bumpy
surface that stirs the air flowing above it. This turbulence
pushes warm air higher and
creates a vertical draft that
cools the forest below. Air flowing through old-growth forests can
reduce summer air temperatures as much as 4.5 degrees Fahrenheit
compared with plantations. Old-forest characteristics such as con-

Another migrant from tropical forests, Vaux's swifts flock in large groups as they
gorge on insects at the top of the forest.

tinuous canopies, high biomass, and complex understories provide a critical refuge in a progressively warming world.[10]

It used to be that the best way to study the canopy was to cut down the tree. This method provided early canopy researchers with a scattered mess of broken pieces that they reassembled, leaf by lichen, into a Frankenstein model of a forest canopy. Why not climb the trees and see what lives there? That was the question two Oregon State University students, Diane Tracy and Diane Nielson, posed to their faculty advisor, Bill Denison, in the 1960s. Both students were experienced mountain climbers, and they had the equipment and the curiosity to explore tree canopies firsthand in repeated, nondestructive ways. The science of canopy studies was launched.[11]

More scientists followed Tracy and Nielson into that world between earth and sky, as canopy researcher Nalini Nadkarni describes the treetops she explores.[12] In 1994, canopy studies in the Douglas-fir region got a boost from the 285-foot construction crane at the Wind River Experimental Forest in southern Washington. Circling six acres of old forest, the crane carried scientists, and the occasional science writer, up and into the undulating canopy of lichen-laden conifers. With access from floor to treetop in a three-dimensional column, ecologists had a movable seat to thoroughly explore the forest for clues about how trees cycle nutrients, how forests affect climate, and what is going on with all those needles. The crane tower was fitted with cameras to record the moment leaf buds opened each spring, and sensors to measure the wafting of vapors between trees and atmosphere.[13]

After sixteen years, researchers came to know more than fourteen thousand individual trees and shrubs at the Wind River canopy site. When research budgets tightened in 2011, the crane was removed. But the tower remains, instrumented to capture the full profile of atmospheric conditions from top to bottom, as part of the National Ecological Observatory Network. There are no secrets at the Wind River tower. Along its height, the tower bris-tles with gadgetry meant to capture and analyze the air breathing in and out of the forest, sap moving through trunks, plant growth, and decomposition, all carefully measured and recorded by automated instruments.

From space, you can tell that Earth is alive, with swirling clouds and sparkling oceans encased in a thin, blue shell of atmosphere. The Wind River tower is one of several hundred AmeriFlux towers across North and South America, dedicated to studying that thin, blue shell. Measuring the flux of carbon, water, and energy moving from forest to atmosphere is essential to the complex accounting necessary for climate-related policies. Entrepreneurs around the world are racing to develop new engineered technologies to suck carbon out of the atmosphere, but here, in the Wind River forest, trillions of needles and leaves already do that through the magic of photosynthesis. Engineered technologies might capture carbon emissions, liquify them, and inject them underground, while nature-based climate solutions focus on carbon-rich ecosystems such as big, old forests, an approach with lower costs, less machinery, and far fewer risks.[14]

At the Andrews Forest, a five-hundred-year-old Douglas-fir is similarly wired, like a patient in a sleep clinic. Equipped with sensors for much of its height, the Discovery Tree is rigged with instruments to measure carbon dioxide concentrations of air every minute of every day, at several intervals along the height of the tree. Instruments pulse with data as intimate as the wetness of its needles and the humidity at its butt flare. A pipe plunges into the ground to eavesdrop on groundwater. Wires snake up the trunk. Several nearby trees wear dendrometer belts cinched at the waist, to measure how much each tree slims down during the day and expands at night as water moves through the trunk. These data hint at the intertwined flows of water, air, nutrients, humidity, temperature, and all the parts that keep the forest humming.

Examining sixty years of watershed data from Westside Douglas-fir forests, geographers Julia Jones and Timothy Perry

found that when old forests are logged, there is an immediate, short-lived pulse in streamflow, particularly during wet seasons, when streams would be running high anyway. After twenty
years, however, the initial pulse of water flowing out of the clear-
cuts slows, and the dense young plantations were using *twice* the
summertime water as unlogged old forests.[15]

    Forests are the birthplace of water. The old forests of the
Douglas-fir region are champions at dealing with lots of water, and
with long periods of almost no water. Precipitation in the western Cascades can be as much as nine feet a year, but only nineteen
inches fall between May and October. However, climate change
has reduced summer precipitation and increased summer temperature, a double-whammy toward reducing summer streamflows.
Jones speculates that old forests use less water than plantations
during hot, dry spells in part because they have many more working parts to hold water. Standing in the cool shade of an old forest at the H. J. Andrews Experimental Forest, Jones points above
her head to layers of mosses, ferns, and lichens that absorb moisture and slowly release it to the forest. Endophytes—the legions of
microorganisms at work inside needles—similarly might draw and
hold moisture in reserve. She gestures toward her feet, where soil
itself, brimming with life, holds moisture. Moving from microorganism to mountain topography, Jones turns toward the ridgeline
four thousand feet above us to point out the architecture of a continuous forest canopy and steep mountain slopes that create a forest air-conditioner. "Sunshine hits the forest canopy, warm air rises
and moves up the slope," she gestures. "Afternoon shade moves
across the slopes, air cools and descends, carrying water vapor that
will condense on the tree tops at night."[16]

The customary mild climate of this temperate rainforest is due in
part to its proximity to the Pacific Ocean. Bathing in marine air
keeps regional temperatures not too hot and not too cool, and it

keeps the rainforest rainy. Prevailing westerly winds thump sodden marine air against the coastal mountains where it rises, cools, and condenses as rain, sometimes snow. The result is some of the heaviest annual precipitation in the continental United States, as much as two hundred inches a year on the western slopes of the Coast Range and Olympic Mountains. Continuing east, marine air drips light rain over millions of umbrellas in Seattle and Portland. Then it thumps again into the Cascade Mountains, where it wrings out much of its remaining moisture, only about half as much as landed on the Coast Range.

But that's what we winkingly call "normal." Our mild climate can be disrupted by an occasional invasion of continental air masses blowing in as dry east winds that strip moisture from forests and set the stage for fire weather. Or by a mid-latitude bomb cyclone that forms when cold polar air dips far enough south to collide with warm tropical air, bringing flash floods, heavy snow, and fierce winds, in what is officially called bombogenesis. Sometimes the undulating jet stream curves far north, inundating the Douglas-fir forest in a warm atmospheric river blown in from the tropical Pacific. Sudden vacation weather in the middle of winter, sometimes called the Pineapple Express, can bring warm air and heavy rain that rapidly melts snow and triggers rain-on-snow floods. Or a high-pressure area can stall between two stationary low-pressure areas and intensify into a heat dome. It used to be said that climate is what you expect; weather is what you get. Lately, our weather, with heat domes, bomb cyclones, roller coaster jet streams, and fire weather, has been more than anyone expected to get.

The most obvious weather to expect in the rainforest is rain. It is raining on us now at the top of the forest canopy. It sounds like applause. A downpour brings an enthusiastic encore as we dip back under the branches. As we descend, the applause sound turns to soft percussion, then to muffled finger-snapping. Farther down, neither the rain itself nor its sound reaches us.

8

You Float Down the Watershed

Streams are bottomless.

—SHERRI JOHNSON, aquatic ecologist

It's raining harder now. The canopy gardens sway in the storm. Farther down, raindrops run a gauntlet of needles, moss, ferns, and lichens, all built to grab passing moisture. Rivulets trickle down the furrows of bark. In the forest, rain doesn't fall; it hangs in the air and saturates you and everything around you. Moss plumps up; ferns unfurl. In old forests, fog sifts through a fine-toothed comb of needles and twigs that provide trillions of platforms for drops to condense.

The ocean, the mountains, and the spin of Earth work together to create a rainy temperate rainforest. Then geology takes over. In the High Cascades, rain and melted snow disappear through cracked and jumbled lava rock; many miles downhill, water pops up in springs. Compared with the perforated lava beneath the High Cascades, the geologic foundations under most of the western Cascades and Coast Range are older and denser. Here, instead of seeping deep underground, water from rain and snowmelt remains near the surface, gathering energy and volume as it cascades down steep slopes. These streams are flashy and mercu-

---

Cold, swift headwater streams shelter the larvae of caddisflies, stoneflies, and coastal giant salamanders, among much else.

rial, quick to respond to drenching rain and quickly melting snow.
As our climate warms, more precipitation in the mountains falls
as rain rather than snow. And snowpacks that historically melted
slowly for months into summer now seem to be melting earlier and
faster.

An Algerian proverb says you can cross a noisy river but not a silent one. Like people, a noisy river announces its shallowness;
the depths and dangers of a silent river are less obvious. In the
Douglas-fir region, small, noisy, roiling streams eventually assem-
ble into full-grown rivers, wide and silent. As a fisheries biologist,
I spent lots of time walking up noisy mountain streams. I plowed
against their determined currents while trying to balance on rolling, slippery cobbles. Each reach of stream had its own marimba
sound. Water dances across riffles in high tenor notes and plunges
into deep baritone pools. Gravel bars sound like xylophones. A big
log can slow the tempo of a stream or stir a cascade fortissimo.
I learned to recognize the changing pitch of water as it flowed toward me across step pools and cobble bars; I could identify a beaver dam before I saw the structure. Water flows through a beaver
dam, so there's no deep plunging sound on the downstream side.
It sounds like the shake of maracas.

Beavers once lived across most of North America, and they
created meadows and wetlands over much of that enormous area.
Those beaver-made meadows generated flat, fertile fields ready-
made for farming and development by American settlers. Beaver
pelts made elegant hats, and secretions from the butt end of beavers were prized by the perfume industry (castoreum imparts notes
of leather, musk, and vanilla). By the early 1900s, trappers had
wiped out 99 percent of the beavers in North America, and settlers
had drained most of the beaver-built wetlands.

Many years before I was slogging up mountain streams, beaver dams were cursed by fisheries managers. They routinely removed the dams to clear passage for spawning salmon. Stream-clearing goes back even farther, to the early days of logging, when rivers were used to transport logs from forest slopes to mills. Loggers cut trees from stream sides and deepened channels to allow a swift flow of logs downstream. As the lowland rainforests were cut down, logging moved farther uphill and upriver. But steep mountain streams were too small to float a load of logs.

The solution was splash dams. Built of logs placed between two permanent buttresses on either side of the stream, splash dams held back enough water to float hundreds of old-growth saw logs. When the barrier bulged with a full load, the dam was blasted open

Having faced near extirpation, beavers are returning to North America in force.

in a sudden torrent of gushing water and logs. Wood-filled floods scoured streams down to bedrock, bulldozed banks, and tore away streamside vegetation for dozens of miles, decimating salmon habitat. If any obstacle survived, workers demolished it with dynamite before the next run of logs. Clearly, these were not ordinary floods. Splash-dam floods carried more debris more often than any naturally occurring disturbance, more comparable to one-hundred-year floods. Biologist Rebecca Miller reported stories of splash-dam floods that "occurred every day with the 5 o'clock whistle on Steel Creek in the Coquille River basin, Oregon." The Coquille River, and more than one hundred other streams in the Coast Range and western Cascades, are still scarred from splash-dam round-ups that occurred between 1880 and 1957.[1]

Even after splash dams were phased out, fisheries managers continued to remove logjams and beaver dams from salmon streams. In fact, Oregon, Washington, and other western states *mandated* the removal of wood from streams after the area was logged. So, when a neighbor told us about a logjam on the Big Elk that needed clearing out, he added, "There's big cedar in it and it's yours if you can get it." We got it and we roofed our house with it. But in the process, we took the roof off the salmon stream.

Salmon evolved with streams overhung by branches and cluttered with wood. Fallen trees create undercut banks and pools where fish hide in deep, cool water. Beaver dams in particular enhance salmon habitat by keeping water tables high and water flowing all summer. Wood in streams develops slippery, nitrogen-rich biofilms, the base of many aquatic food webs. Without beaver dams and wood in streams, the life-giving complexity of rivers dwindled. And so did salmon populations. Scrambling to find reasons for the decline in coastal salmon runs, fisheries biologists took a second look at efforts at stream-clearing. They began to recognize the value of leaving downed logs and branches in streams to create fish habitat, to provide cooling shade, to store nutrient-rich sed-

iment, and to cycle organic matter. In short, to do the work that
beavers had done.

In the 1980s, watershed management in the Douglas-fir region took a 180-degree shift. Managers started helicoptering in giant boulders and built rock-filled wire gabions to recreate eddies

---

Publicly owned forests provide half the spawning and rearing habitat required by
Pacific salmon.

and falls. They hauled in logs to construct riffles and pools, and they secured them in place so they wouldn't wash out in a flood. With imported features subtly arranged to give the illusion of wild nature, streams began to look more like Japanese water gardens than beaver ponds, and it seemed to have little effect on nose-diving salmon populations. In the 1990s, fish biologists took another look at beaver ponds and similar slack waters. They began making their restoration efforts messier and wilder—more like the work of beavers. They allowed the imported logs to shift in the floods, and they allowed beaver dams to stay put. The salmon multiplied.[2]

Now beavers are coming back. Stove-pipe hats are out of fashion and the hydrologic engineering of beaver dams is in high demand. As fish biologists suspected decades ago, the presence of beavers increases populations of juvenile salmon, particularly coho salmon in the Coast Range. When newly hatched coho fry emerge from stream gravels, they quickly move toward slack water, where downed logs provide food, shelter, and protection from predators. Beaver ponds are a ready-made nursery for young salmon, a rich habitat that can temper the effects of a world that is getting hotter and drier. In addition, beaver-built wetlands allow moisture to seep beneath the pond and under surrounding land to create an area of cool, moist vegetation that doesn't readily catch fire. Aerial photographs of burned forests show clusters of surviving green trees where beaver wetlands have created moist oases that resist drought and flames. Once scientists realized how important beavers are to riparian ecosystems, beaver reintroduction became a priority.

Weirdly, however, beavers themselves continue to be trapped in Oregon as pests (even though Oregon calls itself the Beaver State and displays a beaver on its state flag). Beavers' habit of felling trees and building dams can sometimes flood people's yards and damage irrigation systems, orchards, and culverts. Land managers attempt to relocate such problem beavers to places where their natural engineering might be more appreciated, like in the headwaters of salmon and trout streams. In one unlikely restoration effort, biologists in Idaho boxed up suburban beavers and flew them over a remote region called the River of No Return, then dropped boxed beavers by parachutes into the wilderness. Apparently, this was never attempted a second time.[3]

Currently, because there is more need for beaver dams than beavers to meet that need, people have taken to constructing so-called beaver-dam analogs to restore stream systems impoverished by the lack of beaver activity. Government agencies, watershed councils, utility companies, and landowners spend millions of dollars every year to restore wetlands throughout the region, simulating the natural behavior of beavers.

"Streams are bottomless," aquatic ecologist Sherri Johnson tells me. She is referring to the unseen currents of cool water that reach under the bank, under the roots of the adjacent forest, as far as a half-mile from the main stream channel. Between surface and groundwater, this so-called hyporheic zone washes slowly through soil and around roots, leisurely enough to soak up nutrients between stream and forest. These underground waters bubble up in cool springs in the stream's main channel. Many species of bug and fish spend the earliest parts of their lives in the gravels of the hyporheic zone, protected from rising temperatures, hungry predators, and swift currents.[4]

Interactions of forest and stream extend farther than you might expect. Ocean-born nitrogen carried by salmon has been found embedded in temperate rainforest trees. Evidence comes from an isotope of nitrogen, N^{15}, that is relatively abundant in marine algae but rare on land. Yet marine N^{15} has been found in the wood of trees in the coastal forests of British Columbia and southeast Alaska, carried there by salmon. The same thing has been

found in rivers that run through wine-growing parts of California—where there are salmon in the river, there is up to 25 percent more nitrogen in the upland. How?

After spawning, salmon die. Their bodies fertilize streamside vegetation, and scavengers scatter carcasses farther afield. The marine nitrogen isotope is distributed throughout the forest (or vineyard), eventually absorbed by trees (or grape vines) and incorporated into the wood (I don't know about the wine). Comparisons among forested watersheds with differing numbers of spawning salmon show that the amount of this nitrogen isotope found in trees is directly proportional to the abundance of salmon entering the stream.[5]

This nutrient pipeline moves in both directions. Logs from

A dipper bobs up and down as it hunts for bugs in fast-moving streams, often walking underwater in search of food.

the coastal forest wash downstream and into the ocean, where counter-rotating ocean currents sweep logs and other debris into floating reefs that attract a congregation of plankton and fish. Ecologists Jim Sedell and Chris Maser documented over one hundred species of invertebrate and 130 species of fish associated with these drifting habitats. Eventually, the logs sink as the last bit of their terrestrial air is replaced with seawater. Such woodfalls deliver a windfall of food to the ocean floor, where more than forty species of deep-sea invertebrate devour the wood, scattering fecal pellets of forest nutrients across the bottom of the ocean.[6] Driftwood logs from the

Logs washed from the rainforest pile up on the shores of the Salish Sea near Olympic National Park.

temperate rainforest once made up the majority of wood washing onto beaches in places throughout the Pacific. Native Hawai'ians particularly prized Douglas-fir for building their oceangoing double canoes. Other driftwood landed on the treeless Aleutian Islands, where the Unangan people used it to frame their sleek hunting boats made of driftwood and sealskin.

Streams, like shoes, come in different sizes. First-order streams are the smallest perennial waterways; they trickle into larger second- and third-order streams high up in the headwaters. Moving up in size and down in elevation, and cranking up the volume, streams classified as fourth, fifth, and sixth order are considered medium size, and anything larger is called a river. The McKenzie, Lewis, and Wind are seventh-order rivers in the Douglas-fir region; the Amazon (the world's largest river) is a twelve. Scientists think about this network of streams as a river continuum, an ecosystem moving from headwaters to estuary, constantly changing in width, depth, structure, and communities. In this way of thinking, a stream not only flows, it changes.

Let's start our downstream journey at a trickling, first-order stream, like those that make up most of the stream miles within the Douglas-fir region. At the very top of a steep-walled watershed, these headwaters might be a stone's throw over the ridge to the adjacent watershed, and so they offer critical corridors for species moving across the landscape. Choked with branches and leaves, the stream flows in a series of steps. It's shady here, so most of the primary food production comes from invertebrates that shred fallen needles and twigs, and microbes that decompose those shreds, sending nutrients to communities farther downstream.

Soon we join a larger, third-order stream, a lively tumble of stair-stepped pools and turbulent water swirling around big boul-

ders. Here the stream is large enough to shuffle branches and leaves
into heaps along the channel. Sunlight filters in at midday. Invertebrates collect shredded detritus and bits of algae carried downstream. Further down, we plunge into the current of a fifth-order
stream, bobbing through a series of shallow riffles and pools. Here
the stream is wide enough that most of the accumulated wood is
piled up on the sides, and you can see the sky straight up. Photosynthesis cranks up in the sunlight. Various grazers scrape trails
through algae on submerged rocks and wood. You can cross this
noisy stream on loose cobbles, patches of gravel, or a stout log, unless rain, or rain falling on snow, triggers a sudden flood. It's not
only the movement of water, but also the movement of boulders,
trees, and branches surging in a flood that rearranges the structure
of the stream.

Here you might find tailed frogs, among the most primi-
tive of all frog species and made for life exclusively in noisy, swift
mountain streams. Tailed frogs don't croak or make any vocaliza-
tion, and they lack external ear membranes, perhaps because there's
not much to hear above the roar of cascading water. The tail of
this tiny aquatic frog functions like a penis, something that most
other frogs lack. Tailed frogs hold their fertilized eggs internally to
keep a fast-flowing stream from washing them away. Tadpoles hold
tight to slippery rocks for up to four years using their suction-cup
mouths. As adults, tailed frogs can leap, but they cannot land with
any grace; they belly flop on top of their prey.

Farther downstream, a flood has lodged a downed tree
against the bank. Debris collects in the root wad; sediment settles
around sunken limbs. A thicket of red alders has taken hold in a
protected gravel bar. Ambitious young alders and willows are often
the first to reestablish after a big flood, and they are often the first
to get washed out in the next big flood. Older, wiser conifers stand
farther up the slope, away from most torrents. Yet eventually a big

Douglas-fir will lose its grip on the slope and splash down into the stream. Its presence will influence the stream's structure and food webs for centuries.

The underwater community around the fallen tree is filled with food and shelter. Branches that once muffled canopy winds now slow the flow of water. The mesh of submerged branches provides young salmon and trout with what it once provided birds: a labyrinth of hidden resting places. The slowed current drops its load of fine gravel to refresh spawning sites. During my treks up coastal streams, I found the most fish tucked into pools and alcoves sheltered by heavy wood. During winter storms, when riffles turn to roiling floods, fish take refuge behind these submerged logs to avoid getting blasted by rock-churning torrents.

Over time, fungi and microbes digest enough wood for the larvae of beetles, caddisflies, and stoneflies to move in. These grazers sculpt the wood's grain into grooves that feel like slick corduroy. You might find net-spinning caddisflies hidden in these grooves, casting their tiny nets into the gentle current. Borers drill deeper into the log, perforating it with enough oxygen to fuel the engines of decomposition. Mayfly larvae further whittle away the wood surface, and cranefly larvae bore into soft, saturated wood to pupate.

As you float downstream toward the river, the stream gets wider, more open to the sky. The base of the food web changes from decomposing forest debris to photosynthesis. Patchy variations in channel geomorphology, stream flow, water quality, and aquatic life change along the length of forest streams. Throw in a beaver dam or two, and the patchiness increases. What else lives in this ever-changing aquatic community?

When I worked in streams, we used backpack electrofishers to determine which fish lived in the water. The machine sent a

pulse of electricity across a narrow stretch of water to shock whatever was there into a temporary, belly-up surrender long enough for
us to get a head count. Much of that shocking and counting is now
augmented by genomics-based tools that can detect the presence
of aquatic organisms simply by sampling the molecules they cast
in their wake. Everyone sheds their DNA in one way or another.
A scoop of river water contains the genetic fingerprints of animals,
plants, fungi, and even some bacteria that are living upstream.

This environmental DNA (eDNA) allows scientists to discover what species are present, to discern if native or invasive, and
to identify the microbial communities associated with them. And
by searching for genetic clues left over from eating, reproducing, or
shedding, we might glimpse what all these creatures might be doing in there. These new tools, a giant leap beyond "catch and iden-

Harlequin ducks seem to prefer turbulent water, where they breed (in swift forest
streams) and spend the winter (on wind-tossed coastal surf).

tify," help us glimpse the enormous diversity of life interacting in this ecological stew. Using eDNA to examine one stretch of the Alsea River in the Coast Range, researchers were able to classify the DNA they detected into more than nine hundred distinct taxonomic groups. With eDNA, researchers can zero in on a single species of interest, say juvenile coho salmon, and get a snapshot of the world in which the coho live, the hundreds of organisms sharing the same stretch of river—from a single sample of water.

The science of ecology has matured since I started out as an undergrad with a notebook and binoculars. I recall back then there was a certain disdain for old-fashioned, observational natural history. Today, observation-based field research is powered by a modern array of technological tools that greatly increase human senses, including genomic tools of eDNA, acoustic recorders, satellite-based remote sensing, and advances in computation and machine learning to astutely analyze mountains of observations and apply them to conservation and management.

Knowing who lives upstream is important because stream reaches that contain fish are managed with more protections than reaches without fish. Ecologist Brooke Penaluna has used eDNA to identify the highest point in forested streams where fish are present, generally much farther upstream than electrofishing can detect. With eDNA analysis in Washington and Oregon forest streams, she discovered coastal cutthroat trout as much as eight hundred feet farther upstream than previously known in more than half the streams where she sampled waters. "eDNA allows us to see so much more of the life of the stream, life otherwise too small, too rare, or too cryptic to be picked up by older ways of sampling streams," she says. However, Penaluna assures me, the old backpack electrofisher is still useful for collecting physical data that eDNA cannot, including the size, health, and appearance of fish as they lie dazed for a few seconds in your net.[7]

Penaluna's discovery of fish living in so-called non-fish-bearing streams highlights the fact that headwater streams and rivulets make up more than 70 percent of the stream network in much of the temperate rainforest. "Non-fish-bearing," in land manager speak, means there are no salmon to be seen. There are, however, coastal giant salamanders, one of the largest terrestrial salamanders in the world. Adult sallies are elusive, but their clownish, frilly-collared larvae are seen often in these headwater streams, feeding on sculpins or juvenile trout in this "non-fish-bearing" stream.

But salamanders and sculpins don't have the exalted status of salmon. Pacific salmon are an iconic species throughout the temperate rainforest. Salmon connect the entire North Pacific coastline, touching communities from California to Kamchatka. You are standing in Salmon Nation. For at least ten thousand years, people of the temperate rainforest have witnessed the annual return of millions of salmon. Biologist and historian James Lichatowich described the ecological understanding of and cultural respect for salmon that sustained Indigenous salmon fisheries for much of that time. European explorers in the region during the 1790s described a well-developed Indigenous economy based on salmon with specialized technology to capture and preserve fish, and a rich culture with rules to ensure a respectful, sustainable harvest. The Indigenous Pacific salmon industry of the 1790s was probably larger than the New England cod fishery at the same time.[8] It was one of the world's most abundant sources of sustainable protein, comparable to the vast bison herds of the Great Plains. When European Americans entered Salmon Nation, they changed the Indigenous gift economy (the giving of salmon in gratitude) to the Western industrial economy (the taking of salmon for profit). The result over the last 160 years is to have all but lost one of the world's greatest sustainable sources of gratitude and wonder.

In 1976, as a fish biologist with the Oregon Department

of Fish and Wildlife, I witnessed one of the last big commercial salmon harvests on the Oregon coast. I handled thousands of coho and chinook caught by hook and line from a fleet of small fishing boats out of the small fishing town of Port Orford, Oregon. Of course, it wouldn't last. In the 1970s, the Pacific Ocean was an unregulated free-for-all for industrial-scale fishing, and much like the Pacific forests, the "resource" was rapidly being depleted. The crash of Pacific salmon came at the same time as the rampant overharvest of timber in the Douglas-fir region; both forest and fish had been managed as infinitely renewable crops. Hatcheries and plantations replaced wild fish and wild forests in an attempt to improve upon wild nature. It didn't work. By 1991, 214 Pacific salmon stocks were considered at risk under the Endangered Species Act.[9]

Continuing our journey downstream, we are swept into a seventh-order stream (a river) with long runs of boulder gardens between deep pools. Snarls of fallen trees and branches collect gravel at the edge of the current. The riparian forest has more hardwood than in other parts of this rainforest—dogwood, red alder, and big-leaf maple—all shedding carbon into the river. Organic carbon fuels the productivity of these streams. In the Coast Range in particular, where riparian forests are extensive, rivers are the color of Earl Grey tea.

As you float through the estuary toward the mouth of the river, notice deep side channels anchored by sprawling root platforms of Sitka spruce. These so-called spruce swamps provide food and shelter for young salmon on their way to the ocean. Tides wash over the base of the trees and expose long, thin roots that Native people would use to make twine for weaving watertight baskets. Young shoots of Sitka spruce are an excellent source of vitamin C; it is said that Captain James Cook gave his crew spruce beer to help prevent scurvy during their long voyages into the Pa-

cific. However, in the last 150 years, 95 percent of these tidal forests
have been logged and converted to farmland or industry. And recently, mortality of remaining spruces has increased as sea level rise
pushes saltwater farther up these tidal forests.[10]

    This complex movement of water under and through a maze
of wood and rocky obstacles results in water that is high both in
quality and in quantity. Streaming slowly through the forest and
cleansed by natural filters, the river flows down to where it becomes harnessed by humans for drinking, recreation, and indus-

The South Santiam River tumbles over logs and boulders at House Rock Falls,
Willamette National Forest.

trial use. As hydrologist Gordon Grant reminds us, the most valuable resource in these national forests may soon be high-quality
water. "A lot of climate change is about a change in water," he says,
"where it is, what form it takes, when it falls, and where it goes
from there." Relying on national forests for water is not a new idea:
national forests are the largest source of municipal water supply in
the nation. If you had to buy it, that water would cost $7.2 billion
each year. Back in 1897, the Forest Service's founding legislation
cited "securing favorable conditions of water flows" as one of the
missions for the federal forests. In the future, that mission is likely
to be more important than ever.[11]

For now, most mountain water flowing off the western Cascade Mountains is captured behind aging dams built during the
height of twentieth-century boosterism. Recent removals of dams
in the region are showing us that rivers can heal and fish will return, if given a chance. Looking at data from dams recently removed across the United States, Grant and colleagues found that
sediment piled up behind dams dissipated much faster once the
dams were removed than scientists first predicted, sometimes
within days, and anadromous fish that had been blocked from upstream habitat for generations quickly found the way back to their
ancestral streams.[12]

The Elwha River was one of the Olympic Peninsula's most
productive salmon rivers until two dams built in the early twentieth century completely blocked fish passage. After decades of lob-
bying led by the Lower Elwha Klallam tribe, Congress finally
passed the Elwha River Ecosystem and Fisheries Restoration Act,
which allowed the removal of both dams and reopened more than
seventy miles of pristine salmon habitat. In 2023, Native people
welcomed back the first coho salmon run in more than one hundred years.[13]

Throughout much of the twentieth century, people battled
fire, floods, and drought with large-scale, engineered solutions. It

was a time of engineering optimism, when people deeply believed in human mastery over nature. Entrepreneurs redesigned forests, straightened rivers, and built lots of dams. Mid-century booster-ism described Oregon as a treasure-trove of hydroelectricity and led to the construction of seventeen dams on rivers flowing from the western Cascades, and seven more in Washington. These dams were built for many purposes: electricity, flood control, recreation, municipal water, irrigation, and industry. Dams and industrial-scale logging rearranged the landscape on a massive scale. Few people bothered to consider any problems with such colossal re-arrangement—problems such as water pollution, increased wild-fire, and endangered species. Eventually, we would learn the conse-quences of our attempts to bend nature to our will.

More photos, stories
on Mt. St. Helens
Pages B1-4, A16

The Oregonian

Forecast: cloudy;
high, 68; low, 50;
report on Page C8

VOL. 130 — NO. 37,413   SUNRISE EDITION   MONDAY, MAY 19, 1980   52 PAGES   20 CENTS

Eruption decapitates St. Helens; at least 9 die; Spirit Lake gone

By LESLIE L. ZAITZ
of The Oregonian staff

Mount St. Helens exploded in volcanic fury Sunday, unleashing massive mudflows, floods and other land-changing forces that killed at least nine persons, eliminated Washington's Spirit Lake, and sent ash so ash cloud that by Sunday night had moved as far as Wyoming.

The mountain's elevation dropped from 9,677 to 8,400 feet after eruptions began literally ripping the peak apart at 8:31 a.m. Sunday.

The initial explosion was heard more than 150 miles away. The switchboard at the Vancouver, British Columbia, police station "lit up like a Christmas tree" with callers reporting the explosion, a sergeant said.

The intense heat explosion started a 15-mile wide arc around the mountain's north flank, trapping and killing at least eight persons caught in their cars on Washington 504. The explosive force flattened all trees in its path and leveled cabins and resorts.

Authorities also attributed a ninth death to the volcano. They said a coup-dapter was killed when his airplane crashed into power lines near Ellensburg when visibility was impaired by ash.

Hundreds of persons were evacuated from around the mountain, but authorities said many persons were unaccounted for. The Cowlitz County sheriff's department reported the death toll to increase.

The eruptions triggered massive flash floods and lightning. The rain spawned their own forest fires.

Residents of Toutle flee flood

By STEVE JENNING
of The Oregonian staff

TOUTLE, Wash. — The Toutle Lake School was transformed into a helicopter evacuation post Sunday as military aircraft airlifted dozens of residents stranded by the flood-choked Toutle River.

A calamity caused by Mount St. Helens' volcanic eruption sent scores of feet high down the Toutle River Valley. The raging river downed trees, smashed bridges and washed out roads.

Elsie Calvert, who lives in the Toutle River hamlet of Kid Valley, said she knew it was time to leave when a house and several cars floated downstream Sunday morning near her home.

"I said, 'Oh, my God, it's finally happening,'" she said. After landing in a helicopter at the school a few miles west of her home.

"You could hear the trees just roaring," said Mrs. Calvert, who was evacuated with her husband and four children by a U.S. Coast Guard helicopter.

Patrick Gilmore, another Kid Valley resident, flew out in the same helicopter. He brought his pet box constrictor Jonathan, with him. "Our plant (above) eight miles from the school) is still calm but they said something about putting gas," Gilmore said. "I tried to get my pet cat, but trees were blocking the road.

Others said they hadn't seen some neighbors.

BIGGEST SHOW — Mount St. Helens erupts with a vengeance Sunday morning, sending ash over large areas of Central and Eastern Washington.

a ledge on the lake's mouth shore Toutman had refused to leave the area since St. Helens started fuming March 27.

Phil Cogan, Washington Department of Emergency Services spokesman, said there was no sign of Truman's body, which appeared to be buried under an estimated 50 feet of mud and debris.

A Cowlitz County sheriff's deputy said office believes there are a number of campers stranded in the Toutle river drainage area. Their fate was unknown, the deputy said.

The U.S. Geological Survey reported two persons missing from a survey camp at the base of St. Helens. They were Reid Blackburn, about 30, a photographer for The Columbian in Vancouver, Wash. who was working as a freelance photographer for the survey and National Geographic magazine, and David Johnston, about 30, of Menlo Park, Calif., a survey geologist.

Maj. Bill Bowes of the 304th Squadron, said a helicopter crew flew over that camp but saw no one. He said the camp was buried by four to six feet of debris.

The 304th and the Washington National Guard were scheduling additional search and rescue missions for Monday morning.

Helicopter crews rescued about 50 persons who had been dreaded along the Toutle River, including sightseers who had traveled to within 13 miles of Spirit Lake. These sightseers were stranded when the river tore out sections of Washington 504.

Additionally, the Cowlitz County sheriff's department Sunday night evacuated persons east from the Merwin Dam on the Lewis River. Pacific Power & Light Co. reported that mud flowing into its Swift Creek Reservoir had raised the water level by three feet. Leonard Bacon, company spokesman, said the increase provided no hazard to the dam.

The Cowlitz County sheriff's department also began evacuating about 120 families from the Kalama River basin after the National Weather Service posted flash flood warnings for that basin.

Cogan of the state Emergency Services Department said several small bridges on the Lewis River and both forks of the Toutle River had been knocked out by mudflows and fast-moving liquids.

Interstate 5 near Castle Rock was closed several times Sunday as authorities feared floodwaters from the Toutle River could crest over the freeway bridge or damage it. Many motorists opted to utilize U.S. 101 on the Oregon-Washington coast, avoiding a mile-long backup at the Astoria Bridge. Toll booth tenders stopped taking tolls Sunday night to allow the jam to clear.

The U.S. Forest Service reported scores of fires on the Gifford Pinchot National Forest and other timberlands around St. Helens. Firefighting crews were kept out of the area Sunday because of the extreme hazard from volcanic activity, but they were expected to begin battling some fires on Monday

9

You Witness Disturbing Events

An ecosystem, and an ecosystem study, includes disturbances banging all around.

—FRED SWANSON, disturbance ecologist

What you see when you enter a forest is merely a snapshot of its life. There is more to its story. Your whole life's story could not be extrapolated from your high school yearbook photo, or at least I hope not. Think of all the unrecorded events that have shaped who you are; the same is true for a rainforest. Ecosystems, like so much in life, change in an ongoing pageant of life, death, and rebirth acted out at all scales. Some changes occur slowly, like the development of soil or the maturity of old trees. Other changes are lightning fast, like wildfires or landslides. Understanding disturbance ecology changes our picture of the forest from a one-time snapshot to an epic-length movie. And it changes our thoughts about disturbance, from disaster to renaissance.

Few places on Earth face such titanic disturbances as have shaped this part of the Pacific temperate rainforest. These mountains rise from millions of years of volcanic episodes, and this forest has faced millennia of fires, windstorms, floods, and shifting ground. It is a hotheaded place where disturbance is fundamental.

News heralding disaster, such as this 1980 headline about Mount St. Helens, was eventually replaced by the slow return of life, including blossoms of pearly everlasting.

Certainly, people have ignited plenty of disturbances in the rainforest. For thousands of years, Native people used small fires to keep mountain meadows open for harvesting berries and burned valley woodlands to improve hunting and the production of foods such as acorns, camas root, and hazelnuts. When the U.S. government removed Native people from the land, it removed Indigenous burning from the land. After that, disturbance increased dramatically as an influx of American industries replaced forests with plantations, replaced rivers with lakes, and upended ecosystems with invasive species.

For now, let's rewind human history and first consider disturbance as a force of nature wrought by nature. Think of natural disturbance as an unsettling event that pushes you, or a forest, toward a revision that adds a layer of complexity to your growth and development. I doubt you would have made it this far in life without some formative disturbance.

Natural disturbances remind us, as if we need reminding, that humans do not control Earth. This planet, the one that supports all the life we know, is capable of withdrawing that support with a planetary shrug. Disturbance in this region is no surprise. You can read the history of fire in tree rings and the history of floods in layers of sediment, histories as clear as chapters in a family saga. The tricky question is figuring out how we humans fit into that saga in a landscape of frequent disturbance. When people lose their homes, towns, livelihoods, and loved ones, it is coldhearted to suggest that it's just "nature's way." Fires, floods, landslides, and eruptions all extract a costly human toll. On much longer timescales, those same natural disturbances have sculpted the forests we love.

Take fire, for example. One summer years ago, my family and I were camped on the north shore of Waldo Lake near the crest of the Oregon Cascades. A late-afternoon thunderstorm had brought

no rain, just a light sprinkle of ash. Curious, we paddled offshore
to get a better look. A smoke-filled cloud swelled like a bruise over
the ridge beyond our tent. It's odd to think how long we bobbed
on the lake, mesmerized by the rising smoke, until our twelve-year-

---

Wildfire killed most of the trees on this slope above Waldo Lake in the Willamette
National Forest. A second fire reburned the area twenty-five years later.

old son announced, "We need to leave, *now*." He was right. We rushed back to camp, packed up, and paddled in the dark toward the boat launch, propelled by the sound of exploding trees at our backs. Waiting for us at the launch were two Forest Service rangers, unwilling to leave the campground until all their campers were accounted for. By midnight, we were safely settled on the far end of the lake, watching fireworks of sparks and flames as trees detonated over our old campsite. By morning, dense, billowing columns of smoke rose against the summer-blue sky. Helicopters dipped buckets into the lake to douse flames that threatened the north shore campground (a practice now banned in wilderness areas due to potential water contamination). Otherwise, the fire was allowed to burn. It burned ten thousand acres in what is called a stand-replacement fire, killing most of the trees and understory, leaving only a few scattered survivors to reseed the future forest beneath burned snags.

This is not a dramatic wildfire story. This so-called Charlton Fire was not large and was mostly contained within a wilderness area. There have been much larger fires, before and since. Despite its soggy reputation, this temperate rainforest has a wide-ranging pattern of fire frequency: less fire to the north and west, more fire to the south and east. Parts of the moss-draped cedar forests of coastal British Columbia have seen no fire at all for six thousand years; at the dry, southern edge of the Douglas-fir region, parts of the forest naturally burn every twenty to thirty years. In the western Cascades, small lightning-born fires can burn slowly for months until winter rains snuff them out. Or they can blow up into something much larger every century or so. (Along the coast, big, stand-replacing fires have been even more rare.) A few fires, like the one at Waldo Lake, kill all but a few trees; other fires burn lightly through underbrush and never reach crown-killing heights.[1] Of course, fire patterns from the past are not what we can count on with climate change.

Just for a moment, consider forests east of the Cascades, where fire is a very different beast. Those naturally dry forests have evolved to burn frequently, between two and twenty-five years, with low-intensity fires that thin out saplings and undergrowth. The human effort to stamp out these fires over the past century has left naturally dry forests packed with unnaturally high levels of fuel, ready to burn in bigger, hotter wildfires in a climate that is getting hotter and drier. In response, forest managers through-out much of the western United States now reduce overloaded fuels by thinning forests, removing underbrush, and enlisting con-trolled burns. However, such efforts make little difference in the damp, bountiful temperate rainforest. This is an important differ-ence. The exceptional productivity of this rainforest supports some of the world's largest trees and most densely vegetated ecosystems. It naturally contains substantially more biomass than drier, fire-prone forests. Fire management in dry interior forests cannot be assumed to work in a rainforest.[2]

The Westside temperate rainforest has a lot of what *looks* like fuel. Yet downed wood and biologically rich layers of mulch hold prodigious amounts of water. Underbrush is covered by wet blan-kets of moss; older trees are armored by fire-resistant bark. As a re-sult, severe fire has been far less frequent here, averaging once every one hundred to three hundred years, with plenty of time between fires for the rainforest to regrow. Most of the oldest trees in the Oregon Cascades established after high-severity fires during the 1500s. In moist, protected spots in the Washington Cascades and Coast Range, there are trees approaching eight hundred years old or more, having taken root following fire.

Fire scars on living trees throughout the Douglas-fir region testify that mature trees have survived multiple fires and remained productive for centuries. Close examination of these scars reveal many types of fire that have shaped this rainforest through time, including fires intentionally used to clear brush for hunting and harvesting by Native people for thousands of years. Scars record stories of fires that have varied widely in both frequency and severity.[3] Sometimes small fires blow up and burn ferociously when certain conditions collide, such as high temperatures, low humidity, a prolonged dry season, and a hot, dry, gale-force wind blowing in from the high desert. In those conditions, all you need is a spark.

That's what happened on Labor Day 2020. For much of August that year, several small fires had been smoldering on steep slopes of the western Cascades, including a ten-acre blaze deep in the Opal Creek Wilderness, sixty miles east of Salem, Oregon's state capital. Usually, small fires stay small until they are finally doused by winter rains. In 2020, the Pacific Northwest was in the grip of a widespread drought and record-breaking high temperatures; rain was nowhere in the forecast. Instead, a much more dangerous weather system was hurtling down from the Arctic, creating abnormally high pressure east of the Cascades and abnormally low pressure west of the Cascades. That pushed hot, dry winds with hurricane force from the high-pressure Eastside toward the low-pressure Westside, pressing hot air close to the ground, like a lid on a boiling pot.

Images from the 2020 fires look like a disaster movie. Wind blasted across the Cascades and fanned the scattered smoldering fires into ever larger conflagrations. Wind knocked down power lines that sparked more fires. The blaze in the Opal Creek Wilderness blew up to more than one hundred thousand acres *overnight*. Winds up to seventy-five miles per hour swirled burning embers into adjacent canopies as the fire exploded toward the Willamette Valley. Fires merged—Beachie Creek, Lionsgate, and Santiam— while more megafires blew up to the north and south—Riverside, Holiday Farm, Archie Creek—all blown by the same hurricane-force winds, all at the same time.[4]

Firefighting crews came from across the United States and Canada to help contain the firestorms. When all the state and na-

tional firefighting resources were tapped out, volunteers and residents rallied to cut miles of containment lines around rural communities. For days, several megafires swept across the western Cascades, incinerating towns along the Santiam, McKenzie, and North Umpqua rivers and threatening larger cities in the Willamette Valley, including Salem. Smoke filled the region as air quality spiked to hazardous levels, sometimes literally off the scale. Additional fires ignited in Coast Range forests, burning homes and forests within sight of the ocean. In two weeks, the 2020 wildfires burned almost as much of this rainforest as had burned in the previous fifty years. In Oregon alone, the Labor Day fires of 2020 burned more than one million acres and destroyed nearly five thousand homes and businesses. Nine people died in the fires, including George Atiyeh, a long-time Opal Creek resident who had rallied a nationwide effort to protect the old-growth forest.

There have been more wildfires since then. Every few years,

Small wildfires are a natural part of the temperate rainforest, although a strong east wind can whip up a much bigger conflagration.

forest fires are declared to be the worst in recorded history. And they *are*. Very large wildfires can create their own weather. An explosive maelstrom of smoke and flame can rise far above a burning forest to form a pyrocumulonimbus storm cloud, a firestorm. Whirlwinds funnel smoke and heat like a chimney into the stratosphere, sparking lightning and spewing embers in all directions. It's hard to watch your favorite camping spot, not to mention your home or hometown, go up in flames. Every summer, we witness the image in news reports: exhausted firefighters on the front lines, their heavy yellow coats set against a nicotine-stained sky. And every summer, some firefighters lose their lives. You can't put out a megafire.[5] Fire is an element of the landscape that cannot be removed. It's the nature of the beast.[6]

As tragic as it is, windswept fires are not uncommon in the Douglas-fir region. Evidence from buried charcoal suggests that very large fires swept through the region every few hundred years before 1800, resetting large forest stands to early-seral seedlings. The frequency of stand-replacing fires quickened with White settlement in the 1850s and 1860s, when more than one million acres burned in the Oregon Coast Range.[7]

   In 1902, a fire-breathing east wind pushed the Yacolt Burn across a quarter-million acres of the Washington Cascades and killed at least sixty-five people. In 1936, fire whipped by east winds and fed by hillsides of introduced, highly flammable gorse engulfed the coastal town of Bandon, killing ten people and leaving the town in ashes. East winds in 1933 pushed the Tillamook Burn from the Willamette Valley all the way to the Pacific Ocean, torching two hundred thousand acres in twenty hours. Fire doesn't start with wind alone; it needs a spark. In 2017, a fifteen-year-old boy tossed lighted fireworks into a forested canyon thirty miles east of Portland. The Eagle Creek Fire burned for three months and blackened fifty thousand acres. Luckily, no catastrophic east wind pushed the fire into Portland.

In 1910, after horrifying wildfires in northern Idaho and western Montana, the fledgling Forest Service developed a highly efficient firefighting program to attack and control all forest fires. Fires had to be contained by ten o'clock a.m., and all snags in the area over twenty-five feet had to be cut down. After World War II, airplanes were used to fight fires, and Smokey Bear enlisted all of us mid-century campers with the admonishment, "Only YOU can prevent forest fires." Smokey's effort backfired. Without fire to thin them periodically, naturally dry forests throughout western North America grew thick with flammable wood ready to ignite into ever bigger fires. By the twenty-first century, state and federal agencies were spending more than $4 billion each *year* fighting wildfires with aircraft, tankers, bulldozers, flame retardants, and battalions of firefighters.

Fire may have changed the course of human evolution, but most of us don't understand how to live with it. The Forest Service's determination to stamp out all forest fires has encouraged more people to build more homes in fire-prone wildlands, in what is known as the wildland-urban interface, or WUI (pronounced "woo-ee"). Between 1990 and 2020, the number of homes in the United States built at this interface increased by almost 50 percent, along with more roads and power lines. Residents enjoy the tranquil beauty of the forest, and county governments enjoy the additional tax revenue that development brings. But when wildfire sweeps through, it is often the federal government—and federal taxpayers—who bear the cost of firefighting. To protect WUI developments, firefighters cut fire lines by bulldozing, back-burning, and felling unburned trees to keep fire from approaching homes and developments. Sometimes the prevention is worse than the fire.[8]

In the temperate rainforest, the biggest, most severe fires are

blown by east winds of hurricane force that whip flames into an in-
ferno. Like tsunamis and earthquakes, these are low probability
events with profound consequences, difficult to predict, and inher-
ent to this intemperate region.[9] Much more common are mixed-
severity fires that burn in patches, and here the modern landscape
of patchwork plantations may be particularly vulnerable. In the
moist Douglas-fir region, the most flammable tree is a young conifer in a densely stocked tree plantation. Young trees lack the fireproof bark of older trees, their thin trunks and branches dry out
quickly in hot weather, and their dry needles form a continuous
field of highly flammable matchsticks. Several studies have shown
that moderate severity fires in tree plantations burn hotter, travel
more quickly, and consume more soil than fires in adjacent mature
forests.[10] This is particularly worrisome because most tree plantations are located on private land closer to towns than federal forests located at higher elevations.

Now, instead of fighting fire, some land managers are turning
to Indigenous practice to learn how to use fire as a tool to restore
ecosystems rather than fight them. Cultural fires are smaller, cooler
blazes controlled in nighttime and early morning burns, much
gentler than using backfires, logging, and bulldozers to create fire
breaks. These are fires you can walk next to, as Native people have
used them to manage ecosystems for millennia.[11] Tim Ingalsbee, a
fire ecologist and cofounder of Firefighters United for Safety, Eth-
ics, and Ecology, encourages the use of Indigenous cultural fire
combined with fire ecology. He describes the paradox of our current wildfire crisis, as fire suppression has addressed small, short-
term risk and ignored catastrophic long-term risk: "Ecological
integrity should become the new measure for fire management success rather than the number of ignitions successfully attacked."[12]

Over the years, we revisited our Waldo Lake campsite several times to see how the land was faring. The first summer, the ground was burned black and boulders that had been hardly visible in the dense understory before the fire now sat oddly prominent, as if dropped on bare ground the day before. Trees were black, their bare branches curled and drooping. Yet we could see a few green leaves of rhododendron that had emerged from the blackened, ashy ground, sprouting from buried roots. We spotted a small whorl of beargrass; in a few years it would grow into a meadow of nodding white blossoms. Spindly conifer seedlings with a half-dozen tiny green needles were the beginning of a future forest.

This is the way most forests begin in the Douglas-fir region; even the oldest old growth was born from long-ago fire.[13] After several years, blackened bark had sloughed off the dead trees to expose bone-white spires above the green of returning shrubs. Eventually, flowering plants lured in bees and butterflies, in turn pursued by mountain bluebirds. Signs of deer and elk suggested browsing. Perhaps wolves would return?

Instead, fire returned to Waldo Lake in 2022. Again, it started with a lightning strike, and for nearly a month, the fire burned slowly in the wilderness west of the lake. However, unlike the earlier Charlton Fire, fierce east winds blew up in September and exploded the fire across nearly 175 square miles. This so-called Cedar Fire incinerated the burned-over Charlton Fire area, fed by charred wood, pioneering shrubs, and saplings beginning to revegetate.

Such subsequent burns are fairly common. The 1933 Tillamook Burn reburned extensively five times during the next few decades. The 1902 Yacolt Burn reburned fifteen times in fifty years. It seems that severely burned areas are more exposed to heat and drought than in adjacent old-growth forests, and they are more susceptible to invading weeds (including the gorse that kindled the

Bandon Fire). With warmer, drier conditions, we could get more reburns.

I haven't returned to Waldo since the Cedar Fire. I expect it will have changed, again. When wildfires burn through a forest, some areas are severely burned, while others are relatively untouched. This jigsaw pattern makes it possible for surviving trees to provide seeds for areas that were severely burned. But each subsequent reburn can burn up more of the puzzle pieces, removing more seed sources and requiring more time for trees to establish and be ready for the next fire. A wild forest knows how to survive burning and reburning. However, the future forest will have to face the confounding complexities of an altered climate, invading species, ever more distant sources of native seed, and the relentless desire of humans to control nature.[14]

Disturbances can be cumulative. Consider the string of events that began on Columbus Day 1962, when a powerful windstorm slammed into the Pacific Northwest, spawned by a rare typhoon. Wind gusts in Newport, Oregon, hit 138 miles per hour; in Corvallis, gusts measured 127 miles per hour, the same wind force that hit New Orleans during Hurricane Katrina. No comparable storm has been recorded in western Washington or Oregon, before or since. Trees that had stood for centuries blew down across the region. Even worse, clear-cuts had left thousands of miles of forest edges defenseless against the wind. The volume of timber blown down by the storm, seventeen billion board feet, exceeded the Forest Service's *entire* annual harvest for Oregon and Washington at the time. Immediately after the storm, Congress passed special funding to salvage-log thousands of acres and build hundreds of miles of roads to bring wind-thrown timber to sawmills. It was a challenge that Forest Service timber sales met within one year.[15]

However, hastily built roads and cutover hillsides would trigger floods for years to come. On Christmas Day 1964, a cold snap had already brought heavy snow to much of the region when an atmospheric river of warm air blew in from the tropics. The temperature in western Oregon suddenly rose, bringing twenty inches of rain in five days and quickly melting the snow. Surging floods choked rivers and wiped out roads and bridges across five northwestern states.[16]

Ten years later, my husband and I moved to the Coast Range. Our neighbors showed us broken trees and a battered barn that were remnants from those storms a decade earlier. We recycled barnwood to build our house along the upper reaches of

the Yaquina River, a coastal stream that flows through farmlands and large tracts of second-growth forests. The Yaquina has a flashy habit; it rises and falls quickly with rain. Old-timers told us that

Black-backed woodpeckers depend on the burned forest, where they gorge on the larvae of beetles swarming in to feast on fire-killed trees.

our homesite had flooded ten years earlier, so we raised the house on a foundation of three-foot poles. The plan was to allow any floodwater to flow under the house. We'd seen such foundations along coastal Georgia and Louisiana, so it seemed like a good idea.

We were lucky a flood never tested our design. Floods in this temperate rainforest are nothing like rise-and-fall tidal surges in the southern United States. Floods here are not about water. They might be triggered by water—by rain or rain on snow—but the real floods in these mountains are torrents of mud, rock, and broken trees barreling down streambeds at speeds of up to twenty miles an hour. Such battering rams would have buckled our house at the knees. Lucky for us, none ever did.

Anyone who has skied through perfect powder knows that snow holds a varying amount of moisture. In the Douglas-fir forest, snow can hold a lot of water, as much as twelve inches of water in ten feet of snow. That snow, when melting gradually during spring and summer, provides a slow drip of moisture into mountain streams. But when a Pineapple Express blows through, with warm wind and drenching rain, a quickly melting ten-foot snowpack can add a foot of *additional* water rushing down the slopes. In 1996, familiar-sounding weather reports set the stage for another major flood. Four days of heavy rain followed a period of extended cold. Low-level snowpacks released up to ten inches of water in as little as forty-eight hours. Every major river in western Oregon filled beyond flood stage.[17]

Researchers at the H. J. Andrews Experimental Forest were in the catbird seat during the 1996 flood and witnessed massive debris flows barreling down Lookout Creek, a tributary of the McKenzie River, itself a tributary of the Willamette. They watched as more sediment, boulders, and wood entered the river in one *day* than in the previous thirty *years*. Fred Swanson recalled seeing

whole trees surging down the river and hearing the thundering percussion of giant boulders tumbling in the current. Uprooted trees,
thirty inches in diameter and one hundred feet long, piled on top
of the debris flow; a waterfall poured over the road. Gordon Grant
described the entire process as "decades of boredom punctuated by
hours of chaos." Something similar might be said of life on Earth.[18]

Landslide movements might remind you of social movements:
both can be stopped in their tracks by a stand of unbudging elders
or by the failure to attract a critical mass. If landslides manage to
attract a contingent of boulders and broken trees, their increased
mass will increase their momentum. They will rush downslope,
plow through standing trees, and hurl a load of wood, rock, and
sediment into streams and rivers. Where landslides hit stream
channels, they can trigger massive debris flows; in Japan, these are
sometimes called sludge dragons. The largest boulders and logs are
pushed to the snout of the dragon, while the fast-moving slurry of
sediment pushes from the tail. If there is less wood and fewer rocks
at the snout, the sludge dragon races much more rapidly downstream, scouring the streambed.[19]

    Landslides can strike in an instant, even though subtle forces
may have been slowly dislodging the slope for years. Tragically, in
2014, an unstable hillslope near Oso, Washington, collapsed in a
sudden landslide that propelled eighteen million tons of soil, boulders, and broken trees down the Cascade foothills and across the
North Fork Stillaguamish River. As the landslide pushed across
the river, it transformed into a water-saturated sludge dragon driving high-speed slurries of rock and mud and burying a rural neighborhood. It killed forty-three people in two minutes. The Oso
landslide occurred in an area of known landslide activity, which
describes most of the mountainous areas in the Douglas-fir region.
Despite evidence of past landslides on that hillslope, sixty years

of warnings, and a detailed geologic report from 1999 outlining
the high potential for a catastrophic slope failure, county officials
claimed that the Oso landslide (the deadliest in U.S. history) was
completely unexpected.[20]

In a wild forest, occasional large fires can send occasional
pulses of sediment and wood into streams, delivering crucial materials for building complex aquatic habitat. However, massive ad-
ditions of sediment from logging, roadbuilding, and clearing land
can chronically overwhelm streams and destroy aquatic habitat.
Years ago, well-meaning managers pulled wood out of logjams to
clear a path for migrating salmon. Logjams are now regarded as
a biological legacy that benefits aquatic life, including salmon. It
takes those decades of (arguably) boring monitoring to understand
the lasting value of a few hours of chaos.

Parts of those boring decades have been occasionally enliv-
ened by a research gadget at the Andrews Forest: a debris-flow
flume. The flume is built against a steep slope in the forest, like a
snowless Olympic ski jump. Instead of skiers, the flume launches
water-saturated masses of sand and gravel, like those that occasionally propel landslides. Climb the steep stairs to the top and peer
down the 312-foot concrete channel. Imagine the excitement of a
gaggle of geomorphologists down at the far end, gathered to witness the spectacle of debris flowing down the thirty-three-degree
slope. As much as ten cubic yards of sand and gravel are loaded behind steel gates at the top of the flume, sprinkled with water to
saturation. Then the gates fling open and forty tons of slurry spill
down the ski slope. (I was told the engineer who designed the gates
was inspired by the swinging saloon doors in old cowboy movies.)
Video cameras perched along the slide record the movement of debris down the flume and across a grid at the bottom. The surface of
the channel, the coarseness of debris, even the landscape at the bottom can be rearranged to mimic various real landslide possibilities.
A recent experiment added a pool of water at the bottom of the
flume to show how a landslide can produce a tsunami.

In the temperate rainforest, the most destructive floods are
debris flows, and the most destructive debris flows are triggered by
volcanic eruptions. Currently the most threatening volcano in the
region is Mount Rainier, the highest peak in the Cascade Range. A
massive mountain crowned by twenty-five glaciers, Rainier graces
the horizon of the rapidly expanding Seattle-Tacoma area. Beneath
this picture-postcard beauty lies a drowsy giant that has erupted
every few hundred years for the last several thousand. Its next
eruption could produce volcanic ash, lava flows, and floods of in-
tensely hot rock and volcanic gases, called pyroclastic flows. However, the greatest risk might be from cold lahars, rapidly flowing
slurries of mud and boulders driven by swiftly melting glaciers of
snow and ice. Evidence of past lahars lie in deep sediment deposits
from Mount Rainier to the Puget Sound, where more than a million people now live.

Similar to debris flows, avalanches are common in mid to
high elevations of the Cascade and Olympic mountains, where air
temperatures can fluctuate a few degrees above and below freezing.
When deep snow piles up on top of a slick, thawed-and-refrozen
layer, the top snowpack can slide downhill, flattening trees and
anything else in its path. Bare scars from past avalanches are easy
to spot on the face of forested mountains.

An old Douglas-fir shows its age in its broken top, charred bark, or
centuries-old fungus residing in its roots. Such slings and arrows of
outrageous fortune can invite an invasion of Douglas-fir beetles, a
natural part of the conifer forest. The beetles rarely attack healthy
trees; they prefer larger trees already stressed by wind, fire, or root
disease. When female beetles locate a delicious stand of stressed-
out conifers, they send out chemical messages to marshal a mass as-
sault on the weakened trees. Healthy trees defend themselves by
flushing out invading beetles with pitch drawn from resin canals
dispersed throughout the sapwood. Weeping tears of pitch on the

bark indicate where a tree is battling beetles.

A female Douglas-fir beetle carves a tunnel beneath the bark and lays eggs on either side of this gallery; after hatching, the larvae carve their own intricate hallways perpendicular to the gallery as they feed on the phloem. At normal population levels, tree mortality from Douglas-fir beetles is limited to small patches in the forest; the beetle is a symptom, not a cause, of tree death.

Many symptoms accumulate over a Douglas-fir's long life. Dyer's polypore is a fungus that contributes to the condition vividly named butt rot. Trees can live for

Bark beetle females and their larvae carve lacy, tunneled galleries along the inner bark of stressed or dying trees.

centuries with butt rot; the only telltale sign is the pretty, turkey-tail mushroom sought after by crafters who use it to dye fabrics a subtle rainbow of forest-green, yellow, and orange. For the tree, however, this mushroom indicates a fungal infection that slowly weakens the butt of its host until roots can no longer hold the weight of the tree and it falls.

Another fungus, laminated root rot, can cause infections in clusters of trees. The layered sheets of decaying wood, curved and separated at growth rings, look like a book left out in the rain. Douglas-fir and western hemlock are susceptible to this fungus; western redcedar is not. So when the other trees die back, space opens for redcedar to fill. Swiss needle cast is a fungal disease that causes needles to drop prematurely. Native to western North America, the fungus was first discovered in Douglas-fir plantations in Switzerland, thus the common name. The disease has since become a problem along the Pacific Northwest coast, where high-density Douglas-fir plantations have replaced wild Sitka spruce forests.

Trees coevolve with their native pests, often settling into a compatible give and take. But when non-native pests—English ivy, Scotch broom, Russian thistle, Himalayan blackberry—invade the forest, native trees and other species are not equipped to fight. Introduced by good intentions or ignorant disregard, these invasive competitors can put entire ecosystems out of whack, causing changes that fundamentally alter a native landscape. A changing climate can tip the balance further. Insects, being cold-blooded, respond rapidly to warmer weather that speeds up their reproduction and feeding. Even a degree-and-a-half increase in temperature can double the rate of development of bark beetles, leading to rapid population explosions. "Up to now, the Douglas-fir region is the most pest-free region in North America," Jerry Franklin says. "We don't know why. It could be just a matter of time."[21]

Time may be up for Douglas-firs at the driest edges of the region. As we have seen, Douglas-fir is a magnificent and ubiquitous tree species. It is also an opportunist. During the last century, Douglas-fir has expanded into foothills, valleys, and lower-elevation woodlands, especially at the southern ends of the Cascades and Coast Range. Historically, many of these places were ponderosa pine forests or oak woodlands maintained for thousands of years by frequent, low-severity fires. Removing Native fire practitioners and excluding fire from these woodlands have allowed Douglas-firs to spread into places near their lower limit for precipitation and upper limit for temperature, crowding out oaks and pines.

Now, with a changing climate, those harsh site conditions are even hotter and drier, trees are stressed, and insects have moved in. One particular insect, the flatheaded fir borer, is killing Douglas-firs growing on these warm, dry sites where extended drought has drastically reduced available soil moisture. Flatheaded fir borers appear not to bother the pines and oaks, but their effect on Douglas-firs is visible in stands of dead and dying trees in southern Oregon and northern California. People call it Firmageddon.

Dead and dying trees are essential parts of a healthy forest ecosystem. But when the amount of mortality becomes excessive, a buildup of dead trees increases the risk of high-severity wildfires. Most of the recent mortality in southern Oregon has occurred on sites where cultural burning was forcibly stopped, and fire suppression has resulted in increased abundance of Douglas-fir encroaching on native pines and oaks. Add to this a prolonged drought with lower than normal precipitation and higher than normal temperatures, and the stage is set for disturbances beyond what we might have expected.

One change in the forest (fire suppression in fire-adapted forests, say, or the introduction of a new pest) can lead to a rapid chain reaction of consequences, with grim outcomes. Consider the American chestnut, once the dominant tree in eastern North

American forests. Valued for its strong, rot-resistant wood and iconic holiday nuts, the American chestnut succumbed within fifty years to a pathogen introduced from Asia. Something similar happened to the American elm, when an introduced fungus caused a rapid and nearly complete die-off across eastern North America. That could happen here, Franklin warns, and if it does, we don't have a tree species to replace Douglas-fir.[22]

Yet another such loss is happening, again on the East Coast, with the demise of eastern hemlock, a close cousin to our ubiquitous western hemlock. A pest accidently introduced from Japan in 1951, the hemlock woolly adelgid, has spread from north Georgia to coastal Maine, sucking the life out of eastern hemlocks. Scientists who have chronicled the hemlock's decline at the Harvard Forest in western Massachusetts describe the sound of species extinction: "It is the sound of a thousand dying needles falling. It is the sound of a gentle rain." This is the sort of disturbance that no one wants, and no one expects, but it happens when we move species around the world, knowingly or ignorantly.[23]

Now, the so-called acts of God—fires, floods, droughts, and pest infestations—have the undeniable fingerprints of human intervention, accelerated by climate change and pressures from a burgeoning human population. We humans have quite the talent for modifying natural environments to suit our short-term purposes. Earth has an even bigger talent for disrupting our plans, with consequences we can barely imagine.

10

You Find Human Fingerprints

The bureaucratic seas parted and the scientists marched in—things have never been the same since in the region of the northern spotted owl.

—K. NORMAN JOHNSON, forest economist and historian

It was the day after April Fool's 1993, and it was raining in Portland. People lined Martin Luther King, Jr., Boulevard, clutching their opinions like umbrellas as a motorcade pulled up to the convention center. A seemingly endless parade of dignitaries tumbled out of the limousines, like clowns from a circus car. These were no clowns, and this was no circus. The newly elected president of the United States had come to Portland, along with the vice-president, four Cabinet secretaries, a handful of governors, and the head of the Environmental Protection Agency, to mediate the conflict between owls and jobs. A media circus, certainly. The old forests standing on the horizon, their fate in the hands of this entourage, shrugged in mossy indifference.

The ancient temperate rainforest had already seen a lot in its time. It was towering when the first Europeans set foot on the other side of the continent; it was venerable when Lewis and Clark spent a wet winter on the Columbia River. The forest had been born, and born again, from millennia of fires, floods, and volcanic eruptions. But by 1993, more than half of the old forests had been

---

Research in the range of the spotted owl led to a monumental change in the ecological management of federal forests.

cut down, and the people who lined the boulevard had strong opinions about what should be done about it.

Not far away, thousands of people huddled in the drizzle to hear rockers Neil Young, David Crosby, and others belt out a four-hour concert with a message: save the forest. A smaller crowd of loggers and mill workers gathered across town with a different message: save our jobs. And inside the convention center, President Bill Clinton was set to hear testimony from timber industry executives, tribal leaders, union representatives, rural community leaders, and grassroots environmentalists. After listening to more than fifty testimonials from all sides, he saw that the conflict was not about jobs or trees, it was about *change*. "I cannot repeal the laws of change," he said.[1]

The laws of change have dominated this land of fire and rain since time immemorial. People have made their homes here, in the verdant space between volcano and ocean, for at least

Decades of research revealed the importance of dead wood to living forests and streams, such as here at the foot of Madison Falls, Olympic National Park.

twelve thousand years. They were here to witness back-to-back eruptions of Loowit (Mount St. Helens) in the late 1400s. They were here to feel the ashfall from Mount Mazama seventy-seven hundred years ago. And some people were likely here to see the last of the Ice Age floods twelve thousand years ago. Certainly, throughout this time, many people would have experienced the massive forest fires whose scars can still be seen on the rainforest trees. This land was not empty when Columbus set foot in the Bahamas. An estimated sixty million people lived in North and South America in 1491, only slightly less than Europe's estimated population at the time. Long before European explorers entered the Pacific Northwest, villages of extended families were settled into river valleys and coastlines where today's most populous towns and cities have since been established.[2]

The long residency of Native people throughout the region made a lasting impression on the landscape. For at least ten thousand years, they used fire to manage landscapes for food resources and other cultural values. In foothills and valleys, people burned undergrowth to maintain oak woodlands for hunting deer and gathering acorns. Their seasonal burning kept prairies open for harvesting roots of bracken fern and camas. They made seasonal rounds into the Cascade Mountains to collect obsidian from the lava fields and burned meadows for beargrass and huckleberry production. In coastal villages, people burned undergrowth to foster hazel and harvested long strips of western redcedar bark to make fabrics, mats, and baskets. They used cedar planks for building spacious longhouses and decorated their dwellings with carved cedar poles.

From the beginning, this was Salmon Nation. Five species of Pacific salmon filled the rivers for spawning, and people gathered for a few weeks each year to harvest their families' annual supplies of fish. Fishing villages were regional trading centers where dried, smoked salmon was exchanged for bison products from the Great Plains and shells from the Pacific coast, among much else.

Contact with the first Europeans devastated Indigenous

communities by introducing diseases (malaria, smallpox, viral influenza, measles, and more) for which they had no immunity. Millions of Native people died across North America between the late 1700s and the mid-1800s. A pandemic of smallpox that began during the American Revolutionary War spread like wildfire toward the Pacific Northwest and emptied Native villages. When Lewis and Clark entered the Pacific Northwest in 1806, these diseases had already destroyed much of the once vibrant Indigenous society. The Clatsop Indians of the lower Columbia River told William Clark that smallpox had destroyed their people. After thousands of years of history, life for Native people was upended in a few decades. Salmon Nation was silenced.[3]

Less than fifty years after Lewis and Clark mapped the new western American territory, the U.S. government began a national effort to contain Native people on strictly controlled reservations to make way for White settlers. With so many tribes and bands in the rich lands of western Oregon, it was thought to be inefficient to create separate reservations for each group. Instead, in 1855, the U.S. government established the Coast Reservation, where Native people were confederated on one reservation. Initially, the Coast Reservation encompassed 1.1 million acres on the Oregon coast; while it was a huge area, it was less than one-third the size of the people's original homelands. As valuable resources, including oysters, farmland, and timber, lured more American settlers into the territory, the reservation was split up and steadily whittled down. By 1895, the once immense Coast Reservation was gone.

A similar history played out in Washington, following the national policy that showed no regard for the interests of Indigenous people. In British Columbia, the provincial government initially created more reservations with broader access to traditional lands, which helped keep more Native people in places they considered to be their homelands. However, the intent was the same—to contain First Nations and make room for White immigrants.[4]

European interest in the Pacific Northwest was relatively slow to develop. Beginning in the 1500s, Spain opened a Pacific trade route between its colonies in the Philippines and Mexico, and dotted the California coast with missions. Nearly two centuries passed before the next wave of Europeans arrived on the Northwest coast: Russians in search of sea otter pelts; Spanish to secure their claims north of California; and British in search of a Northwest Passage to connect their colonies with the lucrative Chinese trade. It was in this search for the nonexistent passage that British explorers first took note of the rainforest landscape.

What those early explorers noticed most were the *trees*. British explorer George Vancouver described in 1792 the shores of the Salish Sea as "luxuriant" with "a continued forest extending as far north as the eye could reach." Word spread quickly about the giant forests of the Pacific Northwest. Industry investors saw huge amounts of old-growth timber, free for the taking, just in time to feed the California gold rush and San Francisco's meteoric growth. According to historian Stephen Arno, "For the next 100 years, no other industry came close to matching the economic importance to the Northwest coastal region, stretching from northern California to southern British Columbia—an industry featuring old-growth Douglas-fir."[5]

Logging was dangerous work, and the people who worked the woods often lived in makeshift logging camps at the edge of wilderness. The term "skid row" comes from those lumberjack camps of the Pacific Northwest, where skid roads built of wooden planks were used to haul lumber out of the muddy wilderness. By the late nineteenth century, new transcontinental railroads opened the Northwest to an eager timber industry that had cut down the last white pines of the Midwest and were ready to claim "a future of unlimited prosperity" in the forests of the Pacific Northwest.[6]

As the easy-to-reach trees fell, lumber companies built narrow-gauge railways into steeper, more remote areas. By the early twentieth century, the lumber market was glutted, and so only the most valuable trees were logged, leaving smaller trees broken and scattered on the ground—fuel for wayward sparks. Coal-powered locomotives spewing cinders sparked so many wildfires that railroad companies hired lookouts to ride the rails in handcars and put out small blazes where they could. But the lookouts couldn't keep up with the fires.

It was the growing threat of forest fire, and the seeming waste it ravaged on western timberlands, that helped establish the U.S. Forest Service in 1905. Science-based forest management was a new idea in America when the first Forest Service researchers arrived in 1908 to work at Wind River in the southern Washington Cascades. Here they faced massive, old trees, wastefully harvested and frequently burned, in forests that were nothing at all like the cultivated woodlands of Europe or the cutover second growth of eastern North America. Those scientists brought with them the idea that forests could be tamed, protected from fire and wasteful logging, and made to produce sustainable crops of timber. Their successes created new problems in the temperate rainforest.

Shortly after the Forest Service showed up, World War I erupted, and wood products were suddenly in great demand. Sitka spruce was particularly prized for building airplanes that could rival the Germans' powerful air force. Sitka spruce wood is light and strong, with long, tough fibers that do not splinter when hit by bullets. However, a labor strike organized by the Wobblies (the Industrial Workers of the World) had halted logging in western Oregon and Washington, where Sitka spruce grows. Not to be deterred, the U.S. Army formed the Spruce Production Division and developed sixty military logging camps throughout the region to harvest the spruce for the war effort with mind-boggling efficiency. The division lasted less than two years (1917 to 1919), produced 185 million board feet of old-growth spruce lumber, and

left behind a forest of stumps and a network of roads and mills that seeded the region's lumber industry for the rest of the twentieth century. In the words of the U.S. Army at the time: "An army must be sent to make war in the virgin forests, a vast industrial machine must be built up."[7]

A vast industrial machine had already left parts of the forest looking like a war zone when Thornton T. Munger arrived in 1908 to head the fledgling research office at Wind River. Faced with reestablishing a forest amid the ashes of wildfires, and believing that foresters knew better than nature, Munger established a nursery to grow genetically improved seedlings to plant across the burned land. According to the orthodoxy of the day, old trees were worthless and wasteful. Old growth was

Despised by early foresters, standing dead wood has been redeemed as a fundamental element of diversity in the old forest.

described as "decadent," as if old trees were somehow debauched or licentious; plantation forests were "thrifty and efficient," as if more principled and upstanding. "There is little satisfaction in working with a decadent old forest that is past redemption," Munger told a conference of loggers in 1924. He had a particular hatred for standing dead snags: "They stand, fringing the skyline like the teeth of a broken comb, in mute defiance of wind and decay, the dregs of the former forest, useless to civilization and a menace to life of man and of forests."[8]

Wind River was soon designated as an experimental forest focused on replanting burned and cutover land in the Douglas-fir region. Additional experimental forests would be established to pursue research in other forest types: Sitka spruce on the coast; ponderosa pine east of the Cascade crest. In 1948, the Forest Service established the Blue River Experimental Forest in the Oregon Cascades to study the "problem" of old-growth forests, in particular "the conversion of these overmature forests to managed young-growth stands in the most orderly manner with the least delay."[9] The Lookout Creek site at Blue River contained plenty of old trees to study and a complete drainage with several watersheds to compare large-scale harvest methods. Production-oriented scientific forestry was in full swing following Munger's guidelines to convert "degenerate" forests of old growth into "productive" managed stands of second-growth timber.

Other goals soon emerged. A flood on the Columbia River in 1948 wiped out the town of Vanport, Oregon; fifteen people drowned. It was a reminder of the unwillingness of nature to be harnessed, and it gave Horace J. Andrews, the regional forester at Blue River, a reason to worry about the effects of upland logging on downstream water. Andrews directed the new experimental forest to study the effects of timber harvest on hydrology, soil erosion, and fish habitat. With the entire drainage of Lookout Creek contained within its boundary, the new experimental forest was an ideal site for watershed-scale studies. After Andrews was killed

in an automobile accident in 1951, the experimental forest was re-
named in his honor, and watershed research became a signature
part of the H. J. Andrews Experimental Forest.

By mid-century, the vogue in science was turning away from
observation-based field research and toward research in shiny new,
federally funded forest science laboratories. The fate might have
been sealed for a short-lived Andrews Forest, until, in 1964, another reminder of nature's power hit western Oregon and Washington. Sixteen years after the Vanport Flood, another flood re-
awakened an amnesiac public to evidence, once again, of the power
of natural disturbance. The Christmas Flood of 1964 engulfed
two hundred thousand square miles, killed forty-seven people,
and caused more than $540 million in damage across four states.
It surged through the Lookout Creek drainage, washed out roads,
launched landslides, and stirred the curiosity of forest scientists.

There's nothing like a natural calamity to quicken the hearts
of ecologists. Beyond mere rubbernecking at the storm damage,
scientists began considering the role of disturbance in shaping this
steep mountain watershed. The Lookout Creek basin showed itself to be a dynamic system, and the changes that occurred in the
1964 flood resulted from a chain of earlier disturbances, including logging and roadbuilding. The Andrews scientists began to
consider how past events signaled future changes in a brand-new,
disturbance-based ecosystem model.

Meanwhile, science got bigger. By the 1960s, people around
the world began to doubt the optimism of human progress over
nature. They were worried about fallout from nuclear testing, oil
spills fouling coastlines, and dangerous chemicals polluting air and
water. Alarmed by increasing harm to the world's environments,
the National Science Foundation joined an international effort
to channel the power of science and technology to better understand natural systems around the world. The result was the International Biological Program (IBP). The intent was to build models
of the world's major ecosystems to predict the effect of environ-

mental change on ecosystems and on natural resources used for human benefit. The study divided the world into biomes (large areas characterized by similar climates and plant communities) to be dissected, measured, and defined. One subunit of the IBP in North America was the western coniferous biome, the massive evergreen forests of the Pacific Northwest.[10]

In addition to supporting the IBP's lofty intentions, a handful of researchers also saw the organization as a way to keep the Andrews Forest funded. The experimental forest could offer ongoing watershed studies from before and after the recent flood, a patchwork of harvest experiments, and stands of remaining 450-year-old trees. On the strength of their research, Jerry Franklin and forest scientist Richard Waring succeeded in establishing the Andrews Forest as a primary site for the Coniferous Forest Biome Project of the IBP in 1970. In a metaphorical coin flip, the University of Washington was awarded the assignment of investigating commercially interesting young forests, and Oregon State University in Corvallis got the old growth.[11]

Andrews Forest researchers found themselves at a strange intersection: between an effort to understand environmental change in the old-forest ecosystem, and a determined federal policy to exterminate those old forests. Realizing that they didn't understand enough about old growth to know where to begin to build a conceptual model, the researchers set out to learn all they could about this ecosystem before it was gone. Equally providential, the IBP funds came with a mandate that the scientific investigations would be conducted collaboratively in multidisciplinary teams.

Franklin led an eclectic group of scientists into the western Cascades forest, mostly young postdocs—long-haired and curious young men who described themselves as fish squeezers, tree huggers, and bug pickers. They came to the Andrews Forest from a variety of academic backgrounds and soon coalesced into a tight-knit community that shared an excitement for scientific discovery, professional dedication, and joie de vivre. Initially assigned to collect measurements that would feed energy-budget computer models, soon these young researchers were stumbling over dead trees and asking: what's the story here? For the next several decades, all these squeezers, huggers, pickers, and one rock-tapper (Fred Swanson) would throw their monkey wrenches into what we thought we knew about forests.[12]

Science was throwing monkey wrenches into other parts of society, including new science-based legislation that restricted sources of pollution, protected species from extinction, and required public review of proposed environmental impact. Up until the late 1960s, most environmental legislation had focused on aesthetics, setting aside national parks and beautifying America. The 1970s legislation focused on clean air, clean water, a healthy environment, and protection of endangered species. Suddenly, the concern was not so much about the beauty of purple mountain majesty; it was about the degradation of those mountains, and of our own backyards. The laws passed with strong bipartisan support. Perhaps the citizen activism that proved powerful in advancing civil rights and fighting against the Vietnam War emboldened people to confront environmental problems. In any case, the result was monumental. These laws became the foundation of environmental protection in the United States, and they launched the new field of environmental law.

When you focus on pollution, you inevitably move up the pipeline to the source of that pollution. A lot of economic development creates pollution from manufacturing, transportation, energy, logging, and the like, so an attempt to regulate pollution is an attempt to regulate economic development. Up until the 1960s, the economy had been surging, with an increased standard of living for many, but not all, Americans. By the 1970s, that postwar prosperity had begun to stall, with growing unemployment, an energy crisis, and interest rates topping 20 percent. It was easy to blame regulation. Ever since then, we've been fighting over environmental regulation with ever more polarized slogans: jobs versus owls; the environment versus the economy; climate future versus economic future.

By 1978, piles of scientific studies were beginning to merge into a scientific understanding of old-growth forests. Variation was a constant feature: variation in tree size, age, distribution, species, canopy structure, texture, light, and undergrowth. Heavy wood of various stages of decay characterized the forest floor and streams; large live trees and standing dead trees supported a floor-to-ceiling canopy. The complex structure of old forests created an abundance of habitats for myriad organisms. In sharp contrast, the researchers found that dense, rapidly growing conifer plantations were the most *sterile* forest type.[13]

By this time, the Andrews group had fully absorbed the collaborative ethic of the IBP. They were unconcerned with distinctions between Forest Service and university employees, or an individual's rank within those institutions. I remember that researchers in other parts of the country called it the Corvallis effect (although it included researchers from throughout the region), a collegial coalition that is rare among competing organizations, then and now. To keep the spark of creative discovery burning in this corps of scientists, Franklin occasionally organized pulses of energetic fieldwork in locations far beyond the familiar Andrews Forest. Researchers would pack up their scientific tools and camp together for a week, exploring a new, unfamiliar landscape. Each evening, they shared discoveries and hypotheses around the campfire. When Mount St. Helens blew up in 1980, the Andrews team had their intellectual bags already packed.

Early research in the Douglas-fir region had focused on replanting a crop of trees following wildfires or harvests. After the St. Helens eruption, scientists witnessed natural disturbance on an enormous scale and an unexpected amount of variation in the disturbed landscape. Some scraps of the old landscape had managed to survive the eruption, and those became the beginning of a new landscape. Mount St. Helens offered these scientists a once-in-a-lifetime chance to witness the start of a very long-term ecological process and to take a much broader, much longer view of understanding forests.

Perhaps with a similar goal in mind, the National Science Foundation began funding Long-Term Ecological Research (LTER) sites in 1980. The Andrews Forest was one of the first sites selected. "Long-term" referred to a whopping six-year funding cycle, a radical idea in 1980. Even more radical was the expectation to work in partnership with LTER sites across North America, to ask similar questions across multiple sites, and to pursue issues that would require many years—and many funding cycles—to address. The stability of LTER's long-term funding allowed these scientists to ask questions that might take decades or longer to reveal new knowledge.[14]

With "-ologists" of all stripes working shoulder to shoulder, the Andrews scientists looked at the ecosystem from all different angles. They asked each other: what do you see here? The self-named Stream Team saw the conspicuous tangle of logs fallen into streams and questioned how forests and streams were connected. Wood-clogging streams had long been assumed to be a hazard and should be removed; the Stream Team assumed no such thing. They applied a river continuum concept that linked energy inputs at all points along the watershed, from leaf-lined headwaters to log-dominated rivers, to show that streams were important reservoirs of organic matter in the forest, not simply canals for flushing debris out of the watershed.

The investigations of wood in streams prompted other questions. Forest ecologists who had stumbled over all these logs for so many years suddenly saw downed wood as a major contributor to the forest energy budget. A team headed by forest ecologist Mark Harmon questioned whether current logging practices—clear-cutting, removing woody debris, burning slash, clearing brush—were inhibiting the return of nitrogen back to the soil.[15] The question was: how long does nutrient cycling take in a wild forest? The study they proposed would focus on the process of decaying wood and recycling nitrogen over the course of two hundred years. This was an audacious proposal. The funding cycles that pay for most science studies cover work for two or three years, or the shelf life of most graduate students. The Andrews team was pushing more than the budget cycle of research. They were considering the budget cycle of the *forest*.

Up to this time, science had been separated from policy, in part because policymakers didn't think scientists would understand the necessary political compromises, and in part because scientists didn't want those compromises messing with their research findings. The emphasis on interdisciplinary approaches encouraged more collaborations among scientists, managers, and policymakers to address environmental issues together on a bioregional scale. This began a new way to develop science-based environmental policy that would eventually be tested, in the national spotlight, as the owl wars came to a head.[16]

Although much of the groundbreaking research that began with the IBP was focused on forest and stream ecology, there was one wildlife study that broke new ground in science *and* policy. Eric Forsman's pioneering investigation showed that the northern spotted owl depended on old-growth forests, and that the owl's diminishing population was related to the diminishing amount of old forests. This is the owl of the owl wars. Forsman's research became a driving force for years of Congressional hearings, scientific reports, and the subsequent drafting of the Northwest Forest Plan and its strategy to protect the owl *and* the old-growth forests.[17]

Of course, the owl wars were not about owls alone. Clear-cut

logging had become so intensive across the region, with forest and
aquatic habitats so disrupted, that many species of plant and animal were heading toward extinction. The northern spotted owl
landed on the cover of *Time* magazine, but the story also included
marbled murrelets, coastal salmon, red tree voles, and hundreds
more species of the old forest.[18] In a way, the owl wars were the biggest pulse yet for Jerry Franklin and the scientists of the Douglas-
fir region. The what-if questions posed by Congress forced them
to calculate risks to species and ecosystems and to predict harvest levels under each possible permutation of protection. It recalled the multi-science modeling of the IBP, but now much more
was known about old Douglas-fir forests, thanks to a decade of research centered at the Andrews Forest. In this new pulse, scientists
encountered the exotic ecosystems of Congress, the courts, and the
legislative process. They brought their science with them, and with
that, their credibility.

Much hinged on a definition of "old growth." Foresters referred to "large saw timber," and pegged it at eighty to one hundred years, when Douglas-fir trees begin to moderate their youthful, exuberant growth rate and start to pack on wood. The timber
industry tied the definition to something much older—two hundred and fifty or three hundred years—which would allow plenty
of mature and old forest available for cutting if the very oldest old
growth were to become off-limits. Environmentalists, including
Oregon conservationist Andy Kerr, struggled with language, too.
"Old," he said, was not attractive to a youth-worshiping society,
and "growth" sounded like something a surgeon removes. Worse,
"primeval" sounded too much like "prime evil." The environmental
consensus finally settled on "ancient."[19] The legislature, however,
wanted specifics.

Since the days of the IBP, the Andrews scientists had been
developing a more specific definition of "old growth." It included
much of what Munger had abhorred: standing dead trees, downed
dead wood, fallen logs in streams, multiple species and multiple

ages, with complex structures and a multitude of habitats. Their definition described constellations of plants, animals, structures, and functions. Old growth was not a number of years; it was an entire ecosystem.

While federal scientists were refining the meaning of "old growth," federal judges were refining the interpretation of environmental laws. The new laws called for forest plans that protected the viability of species and their habitats in the national forests. But the forest plans emerging in the 1980s were stubbornly focused on producing timber. Meanwhile, lawsuits piled up against the federal forest plans. In 1990, Federal District Court Judge William Dwyer issued an injunction that halted timber sales on federal lands until the agencies could produce credible plans to ensure the northern spotted owl's viability in the context of the whole forest ecosystem.[20] The timber industry in western Washington and Oregon relied on old-growth timber from the federal forests, and it employed tens of thousands of people. An economic recession in the 1980s had put lots of local lumber mills out of business, and new labor-saving technology eliminated lots of jobs. However, many of those workers felt the most threatened by a bird almost no one had ever seen.

The Forest Service brought in wildlife ecologist Jack Ward Thomas to lead an effort to develop a scientifically credible strategy for conservation of the northern spotted owl. Thomas brought in Eric Forsman, along with several other preeminent wildlife scientists, and they presented a plan within six months. However, Congress realized that the issue was larger than owls. It formed the Scientific Panel on Late-Successional Forest Ecosystems, with Thomas, Jerry Franklin, Norm Johnson, and Yale forest scientist John Gordon, plus another 150 or so scientists and resource specialists, and asked them to map all the old-growth stands in the spotted owl region and evaluate alternatives for old-growth forest conservation.

This was to be a science-based plan, so only forest science experts,
not supervisors or managers, were allowed to participate. Rumor
had it that the few forest supervisors who managed to score entry
had to wear orange hunting vests to signify their insignificance. The
timber industry derided the panel as the Gang of Four. The group's
report to Congress translated ecosystem theories into management
practices for the old-growth forests in the Pacific Northwest.

In retrospect, I missed some of the drama of the owl wars because I lived in the Coast Range, where salmon were a bigger deal.
In flashy coastal streams, you can see the effects of land manage-

---

As scientists mapped the forest's structure and function in the range of the spotted
owl, they created a new ecosystem-based approach to managing the nation's forests.

ment directly, because there are few big dams to moderate extreme
flows. Instead, salmon habitat was getting cleared out or smothered by landslides and debris flows, both strongly connected to logging and roadbuilding. Throughout the time that the owl wars
were raging in the Cascades, there was some consensus around
"saving salmon" in the Coast Range.

In 1991, the American Fisheries Society published a bomb-
shell report with a list of 214 salmon and sea-run trout stocks from
the Pacific Northwest that were at risk of extinction. The authors
pinned the decline on an array of problems caused by hydropower,
agriculture, logging, overfishing, and interactions with non-native
hatchery salmon and steelhead.[21] They urged a new ecosystem approach to salmon recovery focused on habitat restoration, structure, and function, entwining the importance of salmon *and* forests to Salmon Nation. It was prescient in that case, when Missouri
Congressman Harold Volkmer reminded the Gang of Four, "And
don't forget about the damn fish!"[22]

With evidence mounting from a growing stack of science reports,
Judge Dwyer would not lift the injunction against timber sales in
the federal forests until the Forest Service produced a scientifically
credible conservation plan for the northern spotted owl. Conflicts
in the Pacific Northwest became a flash point for the presidential
election in 1992. Candidate Bill Clinton promised to hold a "timber summit" if he were elected. He was, and he did. Less than three
months into his first term, President Clinton and his Cabinet-level
entourage convened a listening session in Portland, Oregon. Clinton stressed that his purpose was not to choose between jobs or the
environment, but to recognize the importance of both. At the end
of the day, Clinton gave the forest scientists sixty days to draft a
science-based plan that would be "scientifically sound, ecologically
credible, and legally responsible."[23]

The president's Forest Summit ended on Friday afternoon. The following Monday morning, Jack Ward Thomas assembled one hundred scientists to begin the work. In the spirit of interdisciplinary collaboration begun with the IBP, nurtured by LTER, and demanded by the president, Thomas brought in scientists from an array of backgrounds. They annexed the fourteenth floor of the U.S. National Bank building in downtown Portland to sequester all the working groups of the newly formed Forest Ecosystem Management Assessment Team (FEMAT) for the next two months. They nicknamed their workspace the Pink Tower, and it had a guard stationed at the door to restrict entry to only the science teams. Regular briefings kept President Clinton informed throughout the process.

The work was intense as working groups examined every aspect of the region's old forests. Norm Johnson recalled those heady times in his oral history: "We knew we were on an adventure and knew things were changing. The keys to the kingdom were being handed over to a group of biologists, ecologists, scientists, and everybody else was pushed aside." Johnson felt the weight of the keys in his hands, saying, "I was this guy who had spent his life thinking that sustained-yield would sustain all the forest values." And now Johnson realized that no compromise could provide abundant timber for harvest *and* high levels of habitat protection. Because habitat protection was the law, timber-harvest levels on federal land would have to fall.[24]

What would become the Northwest Forest Plan was unprecedented in its scope and process. Scientists were asked to develop a comprehensive plan for managing vast federal forests for the survival of hundreds of species, and for ample timber to support local economies, on behalf of the White House. (The timeline to deliver a plan was extended from sixty to ninety days as the scientists struggled to provide an option that did both.) In years of retelling the drama of those times, Jack Thomas would sum it up in three

words: *"Obey the law."* He never lost sight of the intent of environmental laws to focus on ecosystems, not on individual species. And
he knew the difficulties of mixing science and policy.[25]

The FEMAT team delivered an option that Judge Dwyer ap-
proved. It led to the Northwest Forest Plan, an enormous blueprint
for managing twenty-five million acres of federal forests within the
range of the northern spotted owl. Its implementation abruptly
shifted the focus of forest management toward conservation of
old-growth forest ecosystems and the streams that flow through
them. The plan recognized more than one thousand species associated with old-growth Douglas-fir forests and placed more than
half the region's federal forest land into reserves. It shifted federal timber production away from harvesting wild forests and toward restoration-thinning of plantations in order to build complex
ecosystems out of former tree farms. And the plan has continued
to guide ecosystem-based management in the national forests for
more than thirty years.

Jerry Franklin summed it up this way: "The Northwest Forest Plan changed everything, but resolved very little."[26] The plan
ended most logging in old-growth forests, despite the efforts of
subsequent administrations to pull out all its teeth. And the plan
protected most of the remaining habitat for the northern spotted
owl, but failed to anticipate the arrival of the invasive barred owl.
The plan promised federal timber harvest of about 1.1 billion board
feet per year, an 80 percent reduction from record-high harvest levels during the previous decade. As it turned out, harvest fell by
more than 90 percent. Over time, the main source of timber sold
from the region's federal forests was from restoration thinning of
overcrowded plantations.

The plan offered new economic opportunities to timber
towns, but effectively helped only those communities with the location and leadership to take advantage of new possibilities. Successful communities engaged in collaborative planning, embraced

adaptability, and leveraged their natural beauty and proximity to
metropolitan markets to build new economic prospects. It left behind more isolated towns with fewer options for revitalization.[27]

The plan's aquatic conservation strategy expanded the scale
of planning from stream reach to watershed and extended riparian buffers to three hundred feet on either side of streams. However, as the federal scientists knew at the time, saving salmon
would require habitat conservation far beyond the borders of the
federal forests, through the lower reaches of rivers, all the way into
the ocean.

President Clinton had said this plan was about change.[28]
Could he have understood how much change was about to happen? For much of the twentieth century, federal foresters had
worked toward attaining a maximum yield of timber. Environmental protection laws complicated that mission, with mandated
attention to multiple uses, endangered species, and public review.
The Northwest Forest Plan shifted the emphasis further away from
timber production and toward the protection of biological diversity, with an emphasis on protecting endangered species and
the old-growth forests on which they depend. The plan began to
coax mature forest characteristics out of a landscape of single-age,
single-species plantations. Perfectly engineered tree farms would
get a messy makeover to include lots of species, ages, and structures, the sort of landscape earlier forest managers had tried hard
to avoid.

Implementation of these lofty goals almost immediately faced major roadblocks. The 1994 midterm elections shifted the power of
Congress toward Republicans, who wanted to limit the author-
ity of President Clinton and defang his signature forest plan. One
quick way was to attach timber-cutting legislation to an unrelated bill and hope that it would pass as a so-called rider. Repub-

lican legislators from Oregon and Washington found the perfect
(and perfectly cynical) legislation on which to attach their rider:
relief funds for victims of the Oklahoma City bombing and ethnic
cleansing in Bosnia. The unrelated rider was audacious: it allowed
timber harvest on federal forests "to the maximum extent feasible
. . . notwithstanding any other provisions of the law." Clinton reluctantly signed the bill into law in July 1995.[29]

The first test of the new rider was to log a Habitat Conservation Area west of Eugene, Oregon. The Warner Creek area had
been set aside to protect the northern spotted owl. Soon after, fire
burned nine thousand acres of that protected habitat. Arson was
suspected. The new rider allowed harvesting the area, presumably because the burned forest no longer provided suitable habitat
for the endangered owl. You see the possibilities: purposely torching habitat could open thousands of acres of protected land to timber harvest. Environmental groups saw the loophole, too, and they
sued. The Oregon District Court upheld the rider that allowed
logging in federal forests and suspended the requirement to comply
with the National Environmental Policy Act, the Endangered Species Act, or any other law that allowed citizen litigation. With no
way to appeal in the courts, frustrated activists turned to wild displays of civil disobedience.

Civil disobedience to protest logging in the Douglas-fir region goes back at least to 1983, in the Siskiyou National Forest in
southern Oregon. Protesters organized by the activist group Earth
First! stood arm-in-arm in the path of road-building machines and
chained themselves to bulldozers to block construction in a roadless area next to the Kalmiopsis Wilderness Area. It took six con-
secutive human blockades and the arrest of forty-four protesters
before a legal victory halted the construction. Later tactics such as
camping in trees marked for harvest brought additional media attention to logging protests in northern California. It was here, in
a privately owned stand of old redwoods, that Julia Butterfly Hill
spent two years camped in the branches of an ancient redwood to

protest a planned clear-cut. While in residence in the treetop, she hosted television reporters and became an "in-tree" correspondent, to broadcast her concern about clear-cutting old growth. Her tactics managed to save only a small grove of redwoods, but it widely publicized the larger loss of old-growth forests.

When the "Salvage Rider" reopened logging in federal forests and suspended the possibility of any legal appeals, protesters deployed all these same tactics in the Douglas-fir region. Rallying the participation of thousands of people and the attention of millions more, protests became a cat-and-mouse game to see who could get to the timber sales first, the bulldozers or the demonstrators blocking the bulldozers. Throughout the federal forests, young protesters buried themselves in logging roads so they couldn't be dragged out of the way; an eighty-year-old woman chained her neck to a log truck.

In the Warner Creek episode, as soon as the district court upheld the salvage logging, protesters headed up to the national forest and built a fort, with a moat, across the only entrance into the timber sale area. A community of protesters occupied this fort for eleven-and-a-half months, through spring and fall rains and heavy winter snow, in a continuous vigil to block the path of the logging crews. Unwilling to cause any more suffering, loggers didn't push past the human blockade. Wary of bad publicity, the Forest Service didn't either. So for nearly a year, these daredevil antics attracted out-of-town tourists, school field trips, local gawkers, and reporters from as far away as the *New York Times*. Eventually, feeling pressure from the Clinton White House, the Forest Service canceled the Warner Creek timber sale and demolished the fort. The protesters were arrested, along with two newspaper reporters on the scene at the time.[30]

Battles continued as subsequent U.S. presidents attempted to rein in environmental laws and Congress continued to press the fed-

eral forests to deliver more timber. One month after the Northwest
Forest Plan was enacted, Jim Furnish became supervisor of the Siuslaw National Forest in Oregon's central Coast Range. "It was really, really bitter times," Furnish said in his oral history. People
within the Forest Service, angry and hurt, felt as if they had done
everything wrong. People in the timber industry felt betrayed.
Even the environmentalists were not sure what would happen.
With morale plummeting within the Siuslaw National Forest and
trust shattered in the community, "we had to completely recraft a
vision of who we were, why we existed, what we were going to do
in the future to be a viable organization and National Forest," Furnish said, "and this was in the midst of losing eighty percent of our
budget and sixty-five percent of our people."[31]

The Siuslaw Forest is a steep forest landscape along the western slopes of the Coast Range. With an abundance of coastal
streams, the Siuslaw has some of the best spawning habitat in the
world for coho salmon and some of the most productive timber-
producing forests. For decades, old-growth timber from the Siuslaw National Forest had kept sawmills running in small towns
along the coast. However, by 1994, after decades of extensive logging and roadbuilding across steep slopes, salmon habitat in the
Siuslaw was in shambles. Concern for protecting streams and
salmon habitat had not been a top priority before the Northwest
Forest Plan. That changed when coastal coho salmon headed toward the endangered species list, followed by the silverspot butterfly on the Siuslaw headlands and the marbled murrelet nesting in
the Siuslaw's last remaining old growth.

As well as creating a protected *space* for butterflies, fish, and
seabirds, the Northwest Forest Plan outlined a protected *process*
for community engagement to help identify shared goals with a
wide range of partners. That meant bringing together entrenched
enemies—timber companies, environmentalists, landowners,
loggers—and helping them find *something* they could agree on.

Reluctantly, the community agreed there was no going back. Furnish took this opportunity to completely transform the mission of
the Siuslaw National Forest, from timber production to ecosystem
restoration.

There was much to restore. A network of logging roads had
been cut into the steep slopes of the Siuslaw, with narrow culverts

Juvenile coho salmon depend on a network of habitats that begins in steep coastal
mountain streams before they enter the ocean.

to channel some of the region's 120 inches of annual rainfall. Two years after the Northwest Forest Plan was enacted, the 1996 flood washed out miles of these logging roads, filled forest streams with sediment, and choked downstream wetlands. Prior to the Northwest Forest Plan, these roads would have been quickly rebuilt, and trees would have been harvested as salvage from streams and landslides. Furnish saw a new path forward in the chaos. He realized that saving salmon, more than anything else, was the concern shared by most people in the Coast Range, but the Siuslaw National Forest couldn't do it alone. The life cycle of Pacific salmon crosses lots of boundaries, from forested headwaters down through agricultural fields, the estuary, and out to sea, far beyond the Forest Service jurisdiction. Saving salmon became a compelling reason for people to work together, across jurisdictional boundaries, to restore the ecological function of forests, streams, and watersheds in the central Coast Range. Their work continues thirty years later.

Free-flowing coastal streams in the temperate rainforest are especially flashy. Red alders growing along the Hoh River are occasionally swept away in floods that rush down from Mount Olympus, in Olympic National Park.

This is the context in which the Northwest Forest Plan came to be. It is regarded as one of the most sweeping changes to forest management in the world. And yet, in addition to the unexpected arrival of barred owls, there were other complications off the radar in 1994. Concern for climate change loomed just over the horizon. Carbon storage, wildfires, invasive species, altered hydrology, and early-successional forest habitats were mostly beyond the scope of policy at the time. But the Northwest Forest Plan was designed to be adaptive. It has been reviewed, scrutinized really, every ten years since its implementation, to identify emerging issues that were peripheral in 1994. The guiding light continues to be ecological integrity and the processes that govern the dynamics of ecosystems.

The plan itself has changed very little since its drafting. It successfully combined a clearer understanding of science and the law. However, no matter how credible and rigorous, scientific findings alone will not resolve political debates, and reducing scientific uncertainty will not reduce political uncertainty. "The Northwest Forest Plan was one of the first science-based, bioregional management plans ever developed for public land, and one of the last," recalled Tom Spies, one of the architects of the plan and its subsequent reviews. "There has been nothing like it since and it seems unlikely that it would happen again."[32] Perhaps managers were frustrated by having to implement a plan they did not draft, and policymakers were frustrated by having to accept uncertainty and reduced expectations. It turns out the keys to the kingdom of management and policy have never again been entrusted so thoroughly to scientists.

Although the NWFP seemed to settle the question of old-growth logging in the range of the northern spotted owl, logging old forests continues beyond the jurisdiction of the plan. The

Tongass National Forest in southeast Alaska contains the largest section of the Pacific temperate rainforest, lying north of our seasonal rainforest zone in what is labeled the perhumid zone. Beyond the reach of the Northwest Forest Plan, the Tongass holds the largest swath of intact coastal temperate rainforest on Earth. Here Douglas-fir gives way to spruce, cedar, and hemlock, and rain overwhelms fire as a characterizing force. Here also, some of the rules covering the rest of the nation's forests have been repeatedly thrown out and reinstated, as the color of federal administrations has switched back and forth between blue and red.

In 2001, the Clinton administration issued the National Roadless Area Conservation Rule to restrict roadbuilding, and by extension large-scale logging and mining, on fifty-eight million acres in the nation's forests. Since then, the Tongass National Forest has been the battleground over the Roadless Rule. Most recently, the Trump administration exempted the entire Tongass from the rule, reopening (again) the forest to timber harvesting and road construction. In 2023, the Biden administration reversed course (again) and restored the application of the Roadless Rule to the Tongass. None of this rule applies to state and private forest, where logging old growth continues.[33]

Changes in tax law prompted more changes in the forest, as large tracts of former private industrial timber lands shifted into the portfolios and tax shelters of private equity firms. Throughout the nation, Timber Investment Management Organizations (TIMOs) and Real Estate Investment Trusts (REITs) have purchased private forests with the expectation of short-term returns similar to other corporate investments. Long-term stewardship of these private lands does not appear to be a priority. Hundreds of thousands of acres of former Coast Range forest have become rows of thin trees harvested every thirty to forty years, long before they reach their peak timber production, and without any chance of becoming a living forest.

The question posed sixty years ago by the IBP was how to predict the ecological effect of disturbances and protect nature's resources for human use. That phrase, "for human use," begs the question: use by which humans, and for how long? The rapid liquidation of private industrial forests is a use that is available to a few currently living humans, and that is paid for by others, many yet to be born. It will take two hundred years or more to replace the wild forest that has been eliminated from the Coast Range, a forest that has been all but destroyed in one human lifetime.

When Jim Furnish came to the Siuslaw in the 1990s, he was appalled by what he saw as the perpetually young, tree-farm structure forced onto the once magnificent coastal rainforest by plantation forestry. "This was *not* the forest that was entrusted to the Forest Service at its founding in 1905," he said. "By eliminating ma-

---

Sitka spruce replaces Douglas-fir on the foggy edge of the Pacific coast and along coastal streams, where it offers critical habitat for marbled murrelets and Pacific salmon.

ture and old-growth stands and replacing healthy, vibrant natural forests with tree farms, the Forest Service failed its public trust."[34]

The National Forest Service grew out of Progressive Era policies of the early 1900s that expected many services from the young agency. It was meant to conserve the nation's forest and water resources, end wasteful logging practices, and reclaim large areas of burned and cutover land. In addition, it was meant to manage public forests as tools of social reform: promote community stability, prevent timber famine, and protect public forests from predatory corporate profiteers.[35] Overarching all these expectations was this statement from the first chief of the Forest Service, Gifford Pinchot: "Where conflicting interests must be reconciled, the question shall always be answered from the standpoint of the greatest good of the greatest number in the long run."[36]

A century later, Jim Furnish hopes his agency can live up to those progressive ideals and fully embrace the mandate of ecosystem management as the greatest good. He sees that the Siuslaw Forest is beginning to grow back into its natural self. Despite the struggles, forests on public land, managed for the public good for the long run, are among the best hope we have for securing viable forest ecosystems into the future.

A hopeful step in that direction was taken in 2022, when the Biden administration issued government-wide guidance for federal agencies to include the Indigenous knowledge of tribal nations in research, policy, and decision-making across the executive branch. In 2023, the U.S. departments of the Interior, Agriculture, and Commerce signed an agreement toward innovative co-stewardship that would provide tribes a greater role in the management of federal lands that hold cultural and natural resources of significance to their Native communities.

Cristina Eisenberg, the Associate Dean of Inclusive Excellence and Director of Tribal Initiatives at the Oregon State University College of Forestry, helped craft the intention of the new federal guidelines. She describes the combination of traditional knowledge with Western science as "Two-Eyed Seeing." As an ecologist and former chief scientist at Earthwatch Institute, Eisenberg oversaw projects with Indigenous communities all over the world. In every community she worked with, in very different cultures, she saw the same approach to nature, an approach that is holistic, reciprocal, long-term, and always changing. "We have to change with it," she said, "and listen to what nature is telling us. Each place at each moment is unique, but the principles are the same."[37]

In many ways, the federal government has made a commitment to manage the nation's land with Two-Eyed Seeing, with both Western science and traditional ecological knowledge informing and inspiring each other. It expands the notion of the greatest good for the greatest number across seven generations into the future. With that, Eisenberg adds reciprocity and cultural humility to achieve the greatest good of participatory, sustainable, peaceful, and just relationships with Earth and each other.

Mature, wild forests still stand on lands managed by the Forest Service and Bureau of Land Management. Environmental laws protecting public land still stand, and new rules promise a more nuanced approach to conservation challenges. For more than a century, federal forest science, management, and policy have changed directions several times as society shifts its priorities and Earth thwarts human-designed plans. But today, our publicly owned forests may be the best chance we can offer for a forest to *be* a forest.

100
80
60
40
20
0

Barred owl

Northern spotted owl

1990   1995   2000   2005   2010   2015

Year

11

YOU BALANCE ON A
RAPIDLY CHANGING PLANET

Preserving 30 to 50 percent of lands for their carbon, biodiversity and water is feasible, effective, and necessary.

—BEVERLY LAW, global change biologist

When we were building our house in the Coast Range, our pioneer neighbors, Charlie and Edith Kruger, took us to see Edith's childhood home on Gopher Creek, several miles and a lifetime away. The nineteenth-century house was gone, as were the meadows where Edith had chased her goats as a child. Instead, there was a forest of massive conifers, thirty inches or more in diameter, towering over a tangle of maple, alder, and a long-forgotten legacy of homestead lilacs and roses. A lot can change in one human lifetime.

A lot has changed in the Pacific Northwest since it emerged from Miocene floods of molten lava and Ice Age floods of melting ice sheets. This land of dramatic disturbances has been the setting for a long history of human residency, rapid subjugation of people and landscapes, paradigm-shattering science, and unimaginable surprises. When Beaver stole fire from a volcano rising from the water, the old trickster probably couldn't imagine how tricky fire would become for people dead set on burning millions of years' worth of fossilized fuel in a geologic heartbeat.

The northern spotted owl that lent its name to this region is being replaced by an interloping relative, the barred owl. The graph shows the percentage of available habitat occupied by each.

A result of humans consuming all that carbon in one fire-hose gulp is illustrated in data collected on Hawai'i's Mauna Loa, the world's tallest mountain in the middle of the world's largest ocean. Since 1958, scientists there have continually recorded levels of carbon dioxide in the atmosphere, and those levels have been climbing steadily. (An eruption in 2023 shut down the observatory for a few months, but monitoring continued at

nearby Mauna Kea.) Nearly seventy years of data make a jagged-edge graph with tiny serrations of winter-to-summer variations set like teeth on an upward bending saw. The CO_2 record goes back even further, in air bubbles trapped in glacial ice. Ice-core records show today's carbon dioxide levels are at their highest point in eight hundred thousand years, warming at a rate ten times faster now than at any time in the past.[1]

A rapidly changing climate threatens the future of humans

The upward bend of a double-handled saw follows the upward trend of carbon dioxide concentrations in the atmosphere (in parts per million). (Data from NOAA Mauna Loa Observatory) and the world to which we are accustomed. Earth itself has seen worse, but we humans have not. We, and the forests we love, have thrived during a ten-thousand-year vacation of relatively stable climate. The mild-mannered climate that nurtured us for millennia cannot be counted on to stay either mild or well-mannered indefinitely. Throughout its 4.5 billion years, Earth has iced over, heated up, dried out, and flooded repeatedly, with massive extinctions of species as collateral damage. These upheavals will happen again.

In 2023, the climate crisis became a financial crisis when several of the largest U.S. insurance companies stopped coverage in parts of the nation because of "rapidly growing catastrophe exposure."[2] Every fraction of a degree of warming comes with more dangerous and costly consequences. If you are reading this book in 2035, you'll look back on today's foot-dragging and denial and wonder, "What were those people thinking!?" Perhaps we were thinking that technology would save us, new technology to solve the problems caused by old technology. There's no shortage of ideas: Space mirrors to deflect sunshine? Sulfur dioxide spewed into the stratosphere? What could possibly go wrong?[3]

Luckily, there is a proven technology for slowing climate change that is far less extravagant: *forests.* Intact forests accumulate massive amounts of carbon. Yet the World Resources Institute reports that, in the United States alone, forests and natural areas are lost at the rate of one football field every thirty seconds.[4] In 2020, cutting down forests around the world released carbon dioxide into the atmosphere equivalent to the annual emissions of 570 million cars, more than double the number of cars on the road in the United States.[5] Measuring our losses in terms of cars and football fields feels strangely appropriate. Since the 1990s, urban and industrial developments have built on top of one million acres of U.S. forests *each year.*

Not to worry, the boosters say. Amid recent escalating goals to reforest the entire planet, the World Economic Forum launched

an initiative to plant a trillion trees across the world by 2030, trees
to offset carbon emissions and replace millions of football fields of
cutover forests. The trouble is, one trillion trees is not a forest. Even
the best-intentioned plans can be a fig leaf. Policies such as the Trillion Trees Act can become a distraction from the more immediate
goals to end emissions from burning fossil fuels and protect existing stores of carbon in *standing* forests. Many of these trillion trees
are destined for monoculture plantations, replacing wild tropical
rainforests with palm-oil plantations. In the United States, timber
industries reassure consumers that three seedlings are planted for
every tree cut down; they don't mention that the new trees will be
cut long before they can provide significant habitat or do much to
tackle climate change. Planting a trillion new trees might help the
planet in a century or two, but saving living forests will help *now,*
when it is of immediate importance to the planet.

I read a poignant op-ed in the *New York Times* recently, en-
titled "I Pledged $1 Million to Plant New Trees. I Wish I Could
Invest the Money in Saving Old Ones." It was written by Roger
Worthington, owner of Worthy Brewing Company in Bend, Oregon. He said he wished the rules of the Forest Service would allow
conservation investors like him to bid on federal timber sales—
not to cut down the trees, but to preserve them. He had pledged
$1 million to reforest a federal forest, yet then became disillusioned
by the seedlings' prospects to do much good as he watched mature
trees being cut down in the same national forest. "Forests on federal land are held in trust for the public. We own them," Worthington wrote. "Shouldn't our elected officials, as prudent trustees,
be erring on the side of leaving strategic forest carbon reserves intact for present and future generations?"[6]

Worthington's plea is echoed by a large contingent of state
and national environmental groups that emphasize the importance
of preserving mature and old-growth forests as immediate and essential climate mitigation. Their motto is: forest defense is climate
defense.

Climate change is tangled up with other crises we face as inhabitants of a finite planet, such as the crisis of species extinction. According to the World Wildlife Fund, wildlife populations around the world declined by 69 percent between 1970 and 2018, and now more than one million species are at risk of going extinct.[7] What's killing the animals? There are any number of reasons: pesticides, poisoned air and water, invading non-native species, and the compounding challenges of climate change. For many species here in the temperate rainforest, the biggest reason is loss of habitat.

Consider the marbled murrelet, a seabird that nests, rather surprisingly, deep in the old-growth forest. To say "nest" is a bit of a stretch. Typically, the female marbled murrelet lays a single egg directly on a moss-covered branch in a big tree, in a forest of big trees, many miles inland from the ocean. It's a lonely childhood for the

A marbled murrelet lays a single egg on a mossy branch in an old-growth forest, hidden from ravens and other predators by a curtain of lichen.

hatchling, which is briefly visited a few times a day by each parent bearing one fish at a time. At fledging, the young bird must find its own way to the ocean and locate a flock of like-minded murrelets.

This life history alone is full of peril. Add in decades of industrial logging that has whittled away mature coastal forests and invited an influx of ravens and jays that scavenge murrelet nests in the remaining few big trees. No wonder marbled murrelets are listed as endangered in Washington, Oregon, and California, where their populations are declining in parallel to the loss of old-forest habitat near the coast. Climate change could reduce the abundance of mosses for nesting and reduce the availability of prey fish in nearshore waters. "The question is whether murrelets can evolve fast enough to survive the additional stress of climate change, given that they are already significantly impacted by human-caused stressors," says Kim Nelson, a wildlife ecologist who has studied marbled murrelets for most of her career.[8]

This nondescript little seabird that few people have ever heard of will never land on the cover of *Time* magazine. Endangered species depend on marketing and promotion. They star in documentaries and fundraising appeals; their images hawk everything from corn flakes to sports teams, including a happily breaching whale that sells insurance. Perhaps these species should earn royalties from the use of their images, a little something to go toward restitution.

Recognizing the crisis of extinction as a crisis of habitat loss, the United Nations Biodiversity Conference adopted the "30×30" conservation plan in 2022, which aims to connect and restore at least 30 percent of the world's ecosystems by 2030. At the start of this race against time, the United States had protected only 13 percent of its federal lands in permanent conservation.[9] There's much work to be done by 2030.

Preserving wild forests could help us address another crisis facing humanity: the scarcity of clean water. Water is essential, of course,

and whether it's too much water, or too little, water affects
our communities, our economy, and our national security.
The U.S. Department of National Intelligence predicted
that by 2040, the global demand for fresh water would
not keep up with supply, unless
we managed the world's water
far better than we do today.[10]
A bevy of new technologies are
being explored, including de-
salination systems, wastewa-
ter bioaugmentation, and that
old chestnut: dams. Again, forests offer a simpler alternative.
Forests slow the rate of runoff from rainfall and lower the
risk of flooding, allowing water to recharge local groundwater systems and filter sediments and nutrients.

   The rainforest itself holds
a lot of water. Weighty old forests have many more absorbent
parts than skinny plantations.
Big, deep-rooted trees lift water up from geologic depths
and redistribute it up through
millions of needles. Layers of mosses, ferns, and lichens absorb

Western sword fern, usually a lush, hardy evergreen understory, has been dying in
small patches near Seattle and elsewhere from mysterious, unconfirmed causes.

some of the region's ample precipitation; massive trunks store water. Even the architecture of a continuous forest canopy keeps air cool enough to condense water vapor into a slow drip. A landscape of bare ground and saplings simply dries out.

More than half the drinking water consumed in the United States comes from forested land. In Portland, Oregon, among other western cities, more than 90 percent of drinking water comes from national forests alone. A lot of that drinking water falls as snow. A foot of mountain snowpack can hold an inch of water in reserve until summer. However, a warming world flips snow to rain and melts the diminished snowpack earlier in the season, with less water available late in the dry season. Scientists say we can expect the annual snowpack in the Cascade Mountains to decline by nearly 75 percent by the end of the century.

Weather is not climate, but extreme weather is increasing around the world, and climate models suggest that more severe weather is in store as the planet warms. Weatherwise, the Pacific Northwest is breaking high-temperature records all the time, and unusual weather events are happening with unusual frequency and intensity. An extreme heat wave hit much of the region in June 2021. I was camping in the Cascades during this so-called heat dome when the temperature hit 102 degrees Fahrenheit near the crest of the mountains, thirty degrees hotter than normal for that time of year. Mind you, the official temperature in Portland reached 116 degrees Fahrenheit, and an unofficial measure of the city's asphalt hit 180 degrees. Tender green foliage on south- and west-facing branches turned brick-red before falling in a strange rain of dead needles. Farther north, the town of Lytton, British Columbia, recorded 121 degrees Fahrenheit that day, the highest temperature in all of Canada's history. The next day a wildfire destroyed Lytton's desiccated downtown; weeks later, the surrounding area was buried by mudslides. The chief coroner of British Columbia reported 619 deaths related to the heat event. Another 255 deaths were attributed to the heat in Oregon and Washington. The World Weather Attribution group, an international team of climatologists, concluded that the intensity of the heat wave was almost certainly linked to human-caused climate change.[11]

Friederike Otto, cofounder of World Weather Attribution, warns us not to blame every disaster on climate change. Disasters occur, she says, when a hazard such as climate change meets vulnerabilities such as degraded ecosystems, poor policies, and social injustices. Habitat loss is still three times more important than climate change in vertebrate extinctions, she says. Climate change has pushed conditions for more frequent fires and opened the conversation about logging the rainforest to protect it from burning. Fire is intrinsic in this forest; the larger vulnerability is logging. Blaming only climate change provides a politically convenient excuse for those responsible for creating vulnerability.[12]

You might think that the temperate rainforest would be sheltered from the worst effects of a hotter, drier world, with a big, cold, wet ocean at its doorstep. However, more than thirty years ago, Jerry Franklin and a team of fellow scientists, in work similar to Otto's, foresaw that in the Pacific Northwest climate change would multiply a landscape's vulnerabilities. Here, they predicted, changes in the region's wild forests would be triggered by disturbance long before climate change alone had time to make a mark. The response of mature forests to climate change was predicted to be gradual, with substantial time lags, because the forest's complex structure and function can moderate climate extremes and because established trees in this rainforest have evolved to accommodate a battery of disturbances throughout their long life spans. However, if the forests are significantly altered by logging or other large-scale disturbances, the researchers predict, response to climate change will be much more profound.

Among the most obvious vulnerabilities in the temperate rainforest is the pattern of clear-cuts and plantations left behind by twentieth-century forest management. As resilient as the wild rainforest might be, much of the region has been redesigned for timber production in the last century, with ecologically diverse native forests converted to intensively managed, single-species plantations. Such areas of relatively young, uniform forests are at higher risk to wildfire, insects, and disease than are more ecologically diverse wild forests.[13] Franklin and colleagues conclude that natural forest ecosystems in the Pacific Northwest are expected to show greater resistance to climate change than intensively managed forests.[14]

Yet changes are likely as a warming climate kicks off a cascade of consequences in the Pacific Northwest. A warming climate delays snowfall, reduces snowpacks, and melts snow earlier, so summer drought starts earlier than it did historically. Earlier droughts and increased summer temperatures lead to increased water stress and insect pest damage, so more trees die and add fuel to the forest. Extended drought and high summer temperatures extend the fire season and increase the risk of more, and more extensive, high-severity fires. So, conservationist Andy Kerr mockingly concludes, the only way to eliminate *forest fires* is to eliminate *forests*.[15]

The Douglas-fir region is characterized by distinct gradients of temperature and precipitation, north to south, west to east, up altitudes, and between slopes. Parts of the forest might evade the worst of climate change effects, as they have for centuries, in shady north slopes and deep canyons with cold air drainages where old conifers have survived for more than eight hundred years. On the other hand, drier parts of the Douglas-fir region may already be at their environmental limits for heat and drought. Foothills, valleys, and much of southwest Oregon are further stressed by the lack of ground fires that has allowed an intrusion of Douglas-firs into native pine and oak woodlands. Removing intruding Douglas-firs and re-introducing low-intensity burning can help restore these dry woodlands. However, the same prescription applied to the wetter parts of the temperate rainforest would undermine many of the functions, structures, and values for which these forests are known.[16] As a mythical forest philosopher said, "Don't be hasty."[17]

Unlike drier western forests that evolved with fires burning through an area every ten to thirty years, the wet temperate rainforest might burn an area every one hundred to three hundred years. The rainforest's prodigious growth and century-long intervals between fires make thinning a futile exercise; underbrush will grow back long before fire is likely to return. Yet state and federal lawmakers have made thinning a priority for limiting wildfires in all parts of the West, including the wet rainforest. The U.S. government budgeted nearly $4.9 billion for Forest Service Wildland Fire Management in 2023, and more for "hazardous fuel reduction."[18] Dry forests throughout the western United States likely benefit from prescribed burning of underbrush and thinning small trees, an expensive analog to how these forests evolved naturally. How do you plan for the inevitability of fire within one of the most productive tree-growing places in the world, a place adapted to fitful fire and tons of rain?

Up until recently, most of the guidelines for managing forest fires have come from dry interior regions that evolved with a lot of low-severity fire. High-severity fires in wet Westside rainforests will require different strategies. Thinning is ineffective in these wet forests, largely because the most destructive fires are driven by high winds (weather) rather than from a buildup of fuel. Fire suppression, anathema to most dry forests, might be a tool to use in wet forests.

Forest ecologists Joshua Halofsky and Daniel Donato are among a group of Pacific Northwest scientists helping to develop realistic strategies for understanding and living with fire in the

rainforest.[19] The key to weathering climate change is to build resilience to natural disturbances, something these wild forests have
been doing for millennia. Extreme events such as large, stand-
replacing wildfires at multicentury scales are inherently rare. They
are more like tornadoes or hurricanes, "black swan" events that
seem too large and too unpredictable to plan for. The best way to
prevent losses from wildfire in the Westside rainforest is to adapt
rural communities to withstand fire. That means: install metal
roofs and soffits; clear buffers around buildings; plan escape routes;
and reduce the number of dispersed developments in wildlands.
Equally important is managing public expectations, because there
is nothing to stop the fury of wind-blown wildfire.

Concern about climate change often focuses on how much carbon dioxide is *released* into the atmosphere from burning fossil fuels. Equally important is how much carbon dioxide is *absorbed* and
stored. Trees are giant carbon storehouses, and trees in the temperate rainforest can hold on to that carbon for centuries. Big trees
with thick trunks and heavy branches hold much more carbon
than smaller trees, and more than half of the total carbon in an old
forest is stored in downed wood and rich forest soil. Yet much of
the carbon stored in trees and soil is lost to the atmosphere during
and after logging. (This is already happening in the Amazonian
rainforest, where a combination of climate warming and deforestation has turned this once immense carbon sink into a carbon
source, emitting more carbon dioxide than it is absorbing.[20])

What about storing carbon in long-lasting wood products?
While it might seem as if lumber could serve as a stable carbon
storage system, no amount of wood construction can match the
carbon storage capacity of a living forest. Researchers found that
only 19 percent of the original carbon in wood will remain in durable wood products, such as buildings and furniture, over a span

of one hundred years (a fraction of a Douglas-fir tree's life span). A whopping 81 percent of the carbon harvested as lumber is emitted to the atmosphere during harvest, processing, or as the lumber festers in landfills.[21]

What about carbon emitted from wildfires? There's no simple answer. About 90 percent of the carbon emitted from a mid-severity fire in the Douglas-fir region remains in the forest as snags and downed wood that will take decades to decompose. Worldwide, wildfires in 2021 released less than 5 percent of total global emissions (or about 1.8 billion tons of carbon dioxide from wildfires, compared with about 38 billion tons from fossil fuels and industry). Of more immediate concern might be the impact of wildfire smoke on human health, particularly the increasing levels of carbon monoxide and small-particle air pollution that seep into buildings far from where fires are actively burning. As wildfires increase in size and intensity, seasonal smoke spreading across the continent could reverse years of improving air quality made possible by the Clean Air Act.[22]

And yet the hazards of woodsmoke are downplayed when wood pellets are burned for power-generation. Governments in Europe and Asia, seeking clean alternatives to fossil fuels, subsidize burning pellets in coal-fired power plants where they release 65 percent *more* carbon than coal per megawatt hour. These are not pellets for your home pellet stove. These wood pellets are manufactured from whole trees, processed into pellets, and shipped overseas, compounding carbon emissions at each step.[23] Certainly forests provide more value when standing than when cut, pressed, and shipped for burning in power plants.

The Pacific temperate rainforest holds more biomass carbon than almost any other ecosystem. These forests are carbon-rich. And with historically long intervals between wildfires and built-in resistance to fire, pests, and diseases, they are likely to survive the challenges of climate change better than most. That is the idea behind a proposal to create strategic forest reserves that would halt logging on fifty million acres of publicly owned mature forests, while leaving younger forests and private tree plantations available for timber. Modeled on the nation's Strategic Petroleum Reserve, which was created after the 1973 oil crisis to guard against future oil-supply disruptions, strategic forest reserves would store carbon as a way to guard against future destabilizing climate shifts.

A team of researchers headed by Beverly Law, a professor of global change biology, set out to identify forests within the United States that could serve as part of this proposed strategic reserve. The most promising forests would have high carbon density, high amounts of critical habitat, high species diversity, and a low chance of mortality from pests, diseases, fire, or drought. Such forests would be primarily on federal land, where government regulation could install permanent protections. If protected, these forests would make strong contributions to biodiversity, water quality, *and* carbon storage throughout the foreseeable future. Not surprisingly, forests with all these attributes are here in the Pacific temperate rainforest, where some of the most carbon-dense forests in the world store carbon in trees for six hundred years or more. Cutting down these big trees, Law says, is like "robbing the Fort Knox of carbon reserves." Forest reserves in these high-priority forests allow the most resilient, carbon-rich forests to persist when the planet needs them most urgently. "Once you harvest them, they're gone," Law says.[24]

The formula of clear-cuts and tree farms is ecological destruction. We've seen the long-term effects of short-term logging in Oregon's Coast Range: jobs dry up centuries before the forest recovers. The best strategies for storing carbon in these high-priority forests, Law told Congress in 2021, is to restrict harvest of older forests on federal land and to lengthen harvest rotations to eighty or more years on private land. If mature intact forests are allowed to grow, they will continue to store carbon, provide critical habitat, and purify air and water in perpetuity. If they are cut, they will release gigatons of carbon into the atmosphere, and it will take a human lifetime or more to reverse the effects of such logging. Keeping trees in the forests where they are already growing is one of the most effective, afford-able ways to slow climate change.[25]

Of course, protecting mature forests alone will not solve all the problems asso-ciated with a changing cli-mate. But preserving mature forests has immediate, pos-itive effects and is less risky than loading the strato-sphere with sulfur. A shift-ing climate is only one of many profound changes oc-curring in the world, and carbon storage is only one of many benefits offered by the temperate rainforest. Let's take a moment here and re-count all the so-called eco-system services we humans

Old photographs of scenes like this are the only remaining evidence of the largest Douglas-firs that once stood in the rainforest at the beginning of the twentieth century.

receive from Earth's wild forests: carbon storage; clean air and water; recharged groundwater; biological diversity; cool, moist buffers against climate change; and a place for recreation, refuge, and spiritual healing. These global services would cost $140 trillion to duplicate, and they are considered "free."[26] And like so many priceless gifts, we tend to assign them no value at all.

Perhaps our changing climate is giving us the opportunity to change our minds about the value of wild forests and other natural landscapes. Think of them as "natural infrastructure" offering many of the same services as engineered infrastructure, often better, cheaper, and more long-lasting. Or think of them as natural experiences offering the world a much-needed sense of connection, wonder, and grace. Or perhaps they are a natural incentive for us to separate the idea of a good and meaningful life from ever-increasing material consumption.

Around the world, wildfires, floods, extreme heat, and ferocious storms are blamed on climate change as if climate change were an act of God. But climate change is an act of hubris and willful neglect. It is the culmination of small decisions made over time by people unwilling, or unable, to consider the larger costs for the longer term. Unlike smoke that blows away when the wind shifts, the use of fossil fuels igniting climate change will not go away without big, conscious human decisions.

When your house catches fire, you think about how much of your life is flammable: birth certificates, family albums, sketchbooks. I thought about all these things as I climbed onto the roof of my house, two stories up, with a limp garden hose in my hand, watering cedar shakes ahead of the flames. Cedar shakes make rich kindling. At the first sign of our house on fire, I had sent my two small children running down the road to the neighbor's house and I called 911. "I'm sorry," the dispatcher said, "we don't cover that area." Word spread in my rural neighborhood, from child to neigh-

bor to the neighborhood post office. There, two loggers heard the news and rushed out to help. They climbed onto the burning roof with me and began ripping off cedar kindling with their felling axes, shake by hand-split shake. We uncovered a beehive between the rafters, burning like a giant candle. The loggers kept ripping and I kept watering, flooding everything below the roof, until at last, the fire was out. Half the cedar roof was gone. That night, from inside the smoke-filled house, I could see clear to the stars.

I could see how things had changed—the loss of some things and the value of what was spared. Eventually, we cleaned up the fire damage and rebuilt the roof with help from neighbors. I never fully thanked the two loggers who saved my house; I never learned their names and I never saw them again. We used our insurance money to help buy a pumper truck for our rural community that lies beyond the reach of 911. You can't repeal the laws of change; life goes on in new ways.

When our roof caught fire, two nearby loggers helped me save the house. The old-growth Douglas-fir rafters did not burn.

12

You Find Ecology in Everything Everywhere

As individuals and as a society, we need to pay attention to the land and rivers—and especially to the consequences of all the things we do.
—JIM LICHATOWICH, fisheries biologist and writer

There's a lightbulb joke for every sector, and mine goes back to my days at the University of Virginia. How many Virginians does it take to change a lightbulb? Ten, of course. One to change the bulb, and nine to say how much they liked the old bulb better. If I learned anything as an undergraduate, I hope it was more than this joke. But for now, the joke works, sort of, as a window into forests.

Individual forests grow, change, and sometimes burn out. The circuitry that binds ecosystems together shifts around all the time, losing connections, gaining connections, switching sides. Forest ecosystems enjoy messing around. Yet we want the places that we love to stay the way we assume they've always been. We can't seem to really grasp that nature is restless, and that our familiar old lightbulb is bound to change.

It seems nature has grown more restless in recent years, accelerated by wildfire, drought, heat, and floods. It's tempting to blame all of this on climate change, when it is human population growth, expanding development, and overheated consumption of natural resources that are fanning the flames. The natural unpredictability

---

Stories about the rainforest ecosystem have been told in many ways.

of ecological change is being pushed toward calamity by decisions we are making as humans.

As a science, ecology is the study of the relationships of organisms with each other and their environments. As a heuristic, ecology ties everything together, with slipknots. An ecosystem is made of slippery bonds between life and environment, with a loose grip to allow things to slide around and change. The science of ecology grew out of naturalists' observations during the hyperobservant nineteenth century. One of the earliest, most revered naturalists at that time was the Prussian explorer Alexander von Humboldt, who traveled through South America without an army or the intention of building an empire. Instead, he carried only scientific instruments and collected only notes about the plants, animals, microbes, Indigenous cultures, and geologic upheavals he saw in an interconnected, shifting web of life. Nothing escaped his notice.[1] This was the work I wanted to do.

However, studying ecology in the 1970s was more like accounting than exploring, like balancing a checkbook using carbon and nitrogen instead of dollars and cents. Back then, ecologists strained against the observational storytelling of their forebears. Instead, we calculated energy exchanges through ecosystems, with feedback loops that cycled back within the closed systems we were trying to track. Using the idea of energy budgets and nutrient cycling, an ecosystem could be defined as being as small as a spacecraft or as large as a planet. As students, we were expected to design studies that were bounded with a limited number of variables that could be reduced mathematically. So, we limited our field sites to an area no bigger than a single-car garage.

My major professor at UVa was Bill Odum; his father, Eugene, had written our textbook, *Fundamentals of Ecology,* and championed our watchword: "An ecosystem is greater than the sum of its parts." Or, as Aristotle seemed to have said much earlier, often misquoted from his treatise *Metaphysics:* things whose parts aren't just piled up in a heap have something going on besides all those parts. Things like communication, relationship, and consequence don't show up in the heap; they emerge as a result of it. Into the economics of systems ecology, Bill Odum added emergence, behavior, and social responsibility. He had us read John McPhee's *Encounters with the Archdruid* and Aldo Leopold's essay "On a Monument to the Pigeon."[2] Ecologists, he insisted, should be *in* the world that they study; they should help solve practical problems and communicate solutions as part of a social feedback loop. For many years after, the unit of value for my wild science would be the "odum."

Perhaps the best part about feedback loops came to me later in graduate school, studying fractals. Branches, ferns, leaves, river deltas, even the neurons in your brain display these infinitely complex patterns created by repeating one simple design over and over in an ongoing feedback loop. It is the basic growth pattern of trees: each branch splitting into ever smaller branches, roots, twigs, leaves, and veins. The psychedelic paisley of fractals decorated the 1970s; neon colors heightened their hallucinogenic effect. But the real beauty was to see slightly imperfect fractal patterns repeated in nature with mathematical patterns far outside the rectangular boxes of Euclidean geometry in our human-designed world. The zigzag patterns of coastlines, tree bark, clouds, and mountains seem random until you take a closer look. Then you see they are made of fundamental shapes repeatedly branching and folding. Large parts resemble smaller parts. As far as you zoom in, it's paisley all the way down. Fractals mapped what seemed like messiness in nature, and fractal geometry made sense of systems that are constantly adapting to their environments.

It turns out those repeated fractal patterns of nature have positive effects on the human brain, measurable by electroenceph-

alograms. When engaging in complex thought, our brains create a
lightning storm of connections that organize spontaneously into
fractal-like patterns of neural networks. And when presented with
nature's fractal images, our brains register high alpha responses,
an indicator of wakeful relaxation and attention. You know
the feeling.[3]

Between mapping energy flows and fractal patterns, I trapped
gypsy moths in the Appalachian forest and mapped songbird territories at Mountain Lake Biological Station. I watched urban development sprawl across the mid-Atlantic coast. Zooming scales and
looping changes created a multifaceted view of the natural world,
as patterns changed with every forest lost or wetland drained. The
more I learned, the more I saw that systems are interrelated, and
changes in one system affect changes in others. Odum was not de-
scribing a static view of nature. His checkbook balancing and feedback loops were meant to demonstrate how systems interact, and
those interactions have consequences, however imperceptible. We
humans are a part of that system, and we affect those feedback
loops. You know this now, but it seemed like a profound idea at
the time.

I migrated to the Pacific Northwest in the mid-1970s, along with
hordes of other newly minted college graduates. The place seemed
like ecotopia. Oregon led the way with laws that restricted urban
sprawl, required bottle deposits, and kept its ocean beaches open to
the public. Oregon governor Tom McCall invited people to visit,
but urged them not to stay. We all visited; we all stayed. We were
part of a wave of young professionals looking for an ecological life.
So, imagine my delight when I found myself in Oregon as a biologist for the Oregon Department of Fish and Wildlife, where there
were wild salmon in the streams and giant trees in the forest. My
husband and I built a little house on the upper Yaquina River in

the Coast Range and marveled at the big fish and wild trees.

I met my neighbors Charlie and Edith Kruger when I was twenty-two and they were in their eighties. They taught me how to grow sweet corn in a rainforest and how to split shakes from a bolt of western redcedar. Charlie had been a logger and a millwright in a lumber mill. He had seen big trees in Oregon before they were cut down, and he had cut down more than a few himself. He described how it took two men (always men) to maneuver the hand-held, twelve-foot, double-ended saw he remembered as a misery whip. First,

Second-growth forests are haunted by old stumps with eyes that once held the springboards of long-ago loggers.

the men cut slit-shaped notches into the tree trunk to hold boards where they could balance above the dense, flared base of the tree. Felling a big Douglas-fir could take a full day or more. Charlie was not a tree-destroyer. He marveled at the big trees, but I never asked him how it felt to see them fall. I didn't know at the time—I guess you never do—that I was sitting at the kitchen table of history, seeing a vanished world through Charlie's eyes.

The chainsaw was introduced to the Oregon woods after Charlie had given up logging for millwork. The chainsaw was the right tool to feed postwar urban development with wild-caught timber from wild, uncut forests. The marvel of ecotopia was short-lived. Every day, giant log trucks rumbled out of the hills, carrying giant trees to the mill. You could hear the machines in the woods; you could smell the smoldering slash-burn piles. And you could see the salmon runs diminishing.

There was very little in my East Coast education that prepared me to understand the scale of commercial harvest of timber and salmon. Working as a fish biologist on the Oregon coast, I counted salmon unloaded from fishing boats; counted their scales for age and growth; counted their tiny, hatchery-coded wire tags; counted the males and females, the juveniles, jacks, and spawners. In summer, I threw nets into estuaries to count smolts heading to sea. In winter, I waded mountain streams to count spawning adults that had survived a gauntlet of obstacles to return, lay eggs, and die. I documented the boom time; I documented the crash.

Discouraged, I turned to another book that Bill Odum had recommended, a book written by an aquatic ecologist who had also witnessed a great diminishment of nature: *Silent Spring,* by Rachel Carson. As a woman in the mid-twentieth century, working for the agency that would become the U.S. Fish and Wildlife Service, Carson was allowed entry into the research world only as a writer in the office of public education. She turned writing and the public understanding of science into a subversive act, a superpower for environmental awareness and political change. She revealed the complexity of cause and effect; she disturbed assumptions; she made the nation care about the loss of species. I followed Carson into writing, trying to make sense of the crash of Pacific salmon that was occurring on my watch.

Of course, Pacific salmon weren't the only species in peril in the temperate rainforest. Beginning in the 1990s, the northern spotted owl and the marbled murrelet were headed toward the endangered species list, along with 214 salmon and steelhead populations, and more to come. The elusive balanced checkbook of nature was way overdrawn. The most obvious link between an owl, a seabird, and an ocean-run fish is the old forest that they each require. The same shortsighted effort to support rural communities by exploiting seemingly unlimited natural resources led to the simultaneous crash of the Pacific Northwest's two longest-standing sources of wealth: salmon and forests.

In the 1840s, Alexander von Humboldt described how organic life is all connected, not in a straight line but in a netlike tangled fabric. In the 1860s, Charles Darwin contemplated a tangled bank of plants, animals, and insects "so different from each other, and dependent upon each other in so complex a manner."[4] In this field trip, we have visited many parts of the forest, tangled parts that illustrate the difference between biology (the study of organisms) and ecology (the study of relationships of organisms to others and to the environment). In ecology, you can't isolate organisms from their environments, or the stream from the forest, or trees from salmon, or plants from fungi. Everything is tangled together in ways that emerge from the heap to surprise and delight us.

Human consciousness is an emergent property of the human brain. No single neuron is responsible for joy or self-awareness. Nonetheless, from all the neurons in the nervous system emerge

human emotions, none of which can be attributed to a single neuron, only to the whole. Human ecosystems are similar. No single person makes the decision to level forests, dry up surface water,
threaten the survival of species, or encourage destructive wildfires.
And yet, here we are.

After college, I kept up with Bill Odum only sporadically, as I was
documenting the collapse of wild salmon in Oregon and he was,
unknown to me, dying in Virginia. I knew that his work with salt
marshes had led him to consider how ecosystems can flip from
one type to another. He had witnessed the slow loss of marshland along the East Coast as a result of small, almost imperceptible human actions. This loss was not planned or desired. Instead,
through hundreds of small decisions, hundreds of small tracts of
marshland were drained for development. Within twenty years,
half the coastal marshes in Connecticut and Massachusetts were
gone. Before he died, Odum published a *BioScience* article enti-
tled "Environmental Degradation and the Tyranny of Small Decisions." He wrote, "Much of the current confusion and distress surrounding environmental issues can be traced to decisions that were
never consciously made, but simply resulted from a series of small
decisions."[5]

    In many ways, we are living now with the tyranny of small
decisions. No explicit decision has been made to cut the last old
tree, dam the last free-flowing river, or drain the last beaver dam.
Yet this is the outcome of incremental actions disconnected from
larger consequences. Eventually, the cumulative effects of small,
seemingly inconsequential decisions trigger the failure of an ecosystem, Odum wrote, and an alternate landscape will take its place.
When it finally happens, the loss will seem inevitable. This process is also called shifting baselines, landscape amnesia, or creeping
normality. It is when a major ecological change, one that we would
not want to happen, is considered perfectly inevitable if it happens

through small, often unnoticeable, increments of change. This is the slippery slope, the boiling frog, the death by a thousand cuts. And it is happening now.

It is easy to condemn the work of past generations when we see how their small decisions have added up to our current colos-

A mountain meadow and a stand of noble firs represent landscapes that are shrinking in area because of changes in climate.

sal problems. They seem stupid, as we will seem stupid to future generations looking back at the decisions we are making now. The tyranny of our own small decisions will constrain the future for generations, particularly our willful ignorance toward species extinction, accelerating climate change, and diminishing forest ecosystems. We think that our wild temperate rainforest has always been here and will always be here. However, we know that within one human lifetime, most of the Coast Range forests have been cut down and redesigned as tree farms. The giant, old trees that graced the forest a century ago are mostly remembered from old photos or haunted stumps, if they are remembered at all. Short-term attention makes long-term changes seem invisible. Consider this: when the Andrews Forest's log decomposition study terminates in 2185, sea level will be twenty feet higher than it is today. New York, Miami, and New Orleans will be underwater, along with much else around the world. By then, the drowned landscape could be mistaken for the way it's always been, and the world we live in now will be as distant as a mythical Atlantis.

The water from this forest stream awakens life, douses fire, cools air, smooths rock, powers the ocean, and makes life possible. It is curious that we choose to flush it down the toilet and buy it back wrapped in plastic.

Taking the long view of forests required generations of scientists working in the Douglas-fir region. They had rejected the idea that these forests should be managed like a crop. They opened our eyes to the complexity of big, muscular trees and the worlds below and above them, all proliferating across the flanks of this steep, unstable, disturbance-driven landscape. They introduced a new way to make environmental policy that put science first. They helped us see the forest as an ecosystem of complex connections and legacies. And much of this new understanding grew from work at the H. J. Andrews Experimental Forest.

In 2023, the Andrews Forest burned in a wildfire that torched twenty-five thousand acres of the Willamette National Forest, sparked by a lightning strike. Many of the older scientists imagined their life's work going up in flames. Graduate students, who had managed to keep their field experiments going through a difficult pandemic, faced starting over from scratch. Forest managers agonized over how they would replace a forest full of research instrumentation that now might be incinerated.

Brooke Penaluna is an aquatic ecologist and lead scientist at the H. J. Andrews Experimental Forest. At the first report of fire, she headed to the forest to collect research records and equipment ahead of the flames. After scrambling to pack the truck with all she could carry, Penaluna took a moment to walk into the smoky forest. "I wanted to stand there among the old trees," she said. "I felt like I was carrying the memories of all the people who had ever worked in this forest, and what this place has meant to all of them. I wanted to see the forest for what might be the last time."[6]

Shock, grief, and a sense of deep loss settled over the Andrews community as the fire continued to burn. Everyone hoped the horrific east winds would stay away. During the long month of burning, as the scientists gathered to hear each incident report issued by fire managers, their conversations began to change. What will be left in the forest? What new questions can we follow as the forest responds? What will be the next chapter of the forest's story? What now?

The east winds did not blow. The Lookout Fire burned through two-thirds of the experimental forest with typical mixed severity; flames hardly touched the paired watersheds from the forest's earliest research, but they tore through Mack Creek, the site of fifty years of aquatic research. Fire skipped the Discovery Tree but burned other old-growth trees as well as plantations. Firefighting crews protected the research buildings, and autumn rain finally extinguished the blaze. At that point, Matt Betts, the lead LTER scientist at the Andrews Forest, had a moment to reflect. He said, "Hope is growing as the forest begins to regenerate, and an explosion of ideas is emerging from the Andrews community about how ecosystems will respond to the fire."[7]

Witnessing such a big disturbance has offered one more interesting plot twist to the story of research at the Andrews Forest. "In several ways we anticipated this fire," Fred Swanson told me. "I am reminded of an old Japanese poem: 'I have always known that at last I would take this road, but yesterday I did not know it would be today.'"[8] Once again, we saw wildfire whipping through the rainforest, sculpting the beautiful places we love with the rough hand of fire. Few places on Earth face such titanic disturbances as those that have shaped this forest of fire and rain. And fewer still manage disturbance with such enduring grace.

Ecology has never been easy for some people to grasp. It seems too vague, without edges where it stops and starts, a bit like old Japanese poetry. You can't draw a line around ecology; you can't fence it in, dissect it, or control it. During the owl wars, lawmakers wanted scientists to deliver a definition of an old-growth ecosystem that could be mapped and posted with "Keep Out" signs. Even today, federal managers of the Douglas-fir region are asked to draw up environmental assessments within each individual timber sale, with little reference to the impacts on a larger, global scale. Such environmental impact assessments are susceptible to the tyranny of small decisions. They are single projects developed with short-term economic goals, and little acknowledgment of long-term effects on air, water, climate, or the legacy we leave for future generations. The result can be a landscape of a thousand cuts.

Worse, of course, is the tyranny of greed and its complete destruction of forests for short-term monetary reward. Federal forests are managed under a ton of paperwork meant to keep managers' decisions accountable to the watching public. Investment companies, accountable only to stockholders, are virtually free to liquidate private forests as if they were any other fungible asset, like oil or cryptocurrency. Ecosystems cannot be traded.

So here we are, standing on a ridge in the Coast Range overlooking miles of Douglas-fir seedlings baking in the summer sun. There is nothing in this vast industrial landscape that hints of the giant forest that stood here a century ago. I am reminded of Aldo Leopold's eulogy to the passenger pigeon, a species that once flew by the hundreds of millions across the Great Plains of North America. Within fifty years, the species was completely exterminated. Leopold wrote, "Because our sorrow is genuine, we are tempted to believe that we had no part in the demise of the pigeon."[9] We are good at showing our sorrow, less so at accepting our responsibility. We want to live in wood houses, while wild forests continue to be cut and species continue to be driven to extinction. This is the tyranny of small imaginations.

As we debate among ourselves about saving the economy or saving the environment, it might be helpful to expand our small imaginations. We live in an ecological world, where the economy is dependent on the environment, not the other way around. With-

out a fully functioning environment, we lack the foundation necessary for real, sustainable prosperity. The 1970s environmental laws
began to codify this common sense: clean up your house or lose it.
Today, these laws can be among our strongest solutions to address
climate change and many more pressing issues—*if* these laws survive. Environmental laws drafted during the 1970s are now more
than fifty years old, pushing retirement age, and we've seen how
easy it is to overturn long-standing legal precedents.[10]

In 2024, the American public was invited to comment on
proposed new rules to protect the nation's mature and old-growth
forests, as part of a strategy to improve the climate resilience of federal forests.[11] The proposed rules amended the Northwest Forest
Plan to address new understandings of climate change and worries
about wildfire. But the proposal did not ban logging mature forests, even though mature and old trees are the most fire-resistant
trees in the forest. And a commitment to old growth makes no
sense without mature forests continuing to grow toward old-
growth complexity.

Forty years ago, Odum called for durable, sustainable policies with long-term environmental goals to avoid the tyranny of
small decisions. He saw how political pressures can force small,
short-term decisions to deal with immediate, specific problems.
He urged policymakers, managers, and scientists to adopt a large-
scale perspective that encompasses the effects of all their little decisions. To do otherwise is to succumb to the tyranny of small
imaginations.[12]

Six years before Lewis and Clark trekked across North America,
Alexander von Humboldt explored South America. He arrived in
northern Venezuela in 1800, during the spring rainy season, where
he was surprised to see the region gripped in drought. From conversations he had with Native people, von Humboldt learned that

the original forest had been clear-cut and replaced with farms, and land with regular rainfall had become a desert. Von Humboldt saw that the forest was necessary for rainfall, and that improper colonial use of the land had long-term ecological and social consequences.[13] To von Humboldt, the environment was not a stable platform for human activity, but a shifting result of ongoing change and consequence.[14]

As you step from the clear-cut back into the old forest, you feel like Dorothy stepping from Kansas into Oz. It is emerald-green and jam-packed with verdant life. You can feel the difference. Japanese researchers have wired up volunteers, sent them out for a walk, and recorded their physiological responses. Compared to urban walkers, those who walked in the forest experienced lowered blood pressure and reduced inflammation. Stanford University researchers took the experiment a step farther, using fMRI technology to scan the brains of participants who walked in forested versus urban settings. The brains of those who spent time in forests showed reduced disturbance activity in the subgenual prefrontal cortex, the area where our moods are regulated. For those willing to immerse themselves more deeply, other studies document a whopping 50 percent increase in cognitive problem-solving by participants after several days they spent backpacking in wilderness. The researchers conclude, "These results suggest that accessible natural areas may be vital for mental health in our rapidly urbanizing world."[15]

This research doesn't so much describe what it means to be a forest as it hints at what it means to be human. More than half the people on Earth now live in urban areas, concentrated in neighborhoods with little access to wild forests or open space. University of Washington psychologist Peter Kahn describes how children accept the environment they live in as the baseline of what is normal, no matter how altered or polluted. In that way, we pass on an envi-

ronmental amnesia, as the amount of environmental degradation increases and each generation perceives the degraded world it inherits as normal, the way it has always been.[16]

If we hope to do better than to give the wreckage of our plunder to the future, then we must be able to see the difference between a forest ecosystem and a tree farm. Every scientist I interviewed spoke of the wild forest's beauty, renewal, and complexity. They described it as a place full of surprises, a place of wonder. Forests are part of what it means to be human, ecologically and spiritually. This field trip was not meant to dissect the parts of the whole, but to celebrate the complexity of the complete whole, the whole tangled forest ecosystem, with all its emergent connections,

processes, and diversity. By putting yourself *in* a forest, from roots to crown, you begin to understand what it means to *be* a forest. And you begin to care.

And so, I have one request of you: *Pay attention.* Stop glancing at your screens; stop multitasking; stop giving in to the tyranny of small decisions or to the ease of not caring. Take heart in what it means to be a forest. Pay attention when you enter this irreplaceable forest ecosystem. Bring your imagination—it is your most powerful tool. As we face crises of environmental loss and de-

---

Before becoming ancient, middle-aged Douglas-fir forests have complex understories and trunks like massive columns.

graded human mental health, we need you to imagine a new way to be in the forest and to recognize your place in the world.

The forest of fire and rain is a paradox. Nestled between a ridge of volcanoes and the grinding edge of tectonic plates, this is a forest of life and death, veneration and consumption, familiar and mysterious, ancient and ephemeral. It is an ecosystem nourished by geologic upheavals and delicate threads of mycelium. It lives off buried treasure; it outlives nations. It is constantly changing.

So, shall we worship our forests, or consume them? How much can we cut before we have cut too much? How much forest do we leave untouched, sanctified, or neglected? These questions hang in the air as we take one last side trip back to my old house in the Coast Range. Like Edith's childhood home, this house sits in a changed world. The house itself is still there, with a new roof and a new family living under it, and a pasture full of goats. The mill town is still there, with its small, family-run sawmill still milling boards, but nowadays those boards are milled from smaller trees cut in plantation-thinning operations. The big paper mill has been redesigned as a high-tech facility for recycling materials from big construction sites. The riverfront has a new marina and a new industry repairing fishing boats for the Pacific fleet—a fleet that is sustained under new, stricter catch limits. Cutover forestland is growing back; filled-in estuaries are carving new channels. Change starts slow, and surprises happen. A forest can burn. A new community can take root. This town has begun a new chapter beyond the false choice of either saving the economy or saving the environment.

The forest we have visited in this field trip exists in fragments. Only in your imagination can you see it as a whole, fully functioning ecosystem across the Pacific Northwest landscape. I recall the wild forest Edith and Charlie described from their pioneer childhoods, and I imagine the giant trees that could be here again, perhaps within *my* grandchildren's lifetimes, if we allow this forest to be a forest ecosystem. It matters what we leave behind. What will you leave behind?

It's raining again as you emerge from the forest, with muddy boots and far more than the sum of your parts. You have imagined the last tree falling with no one around to hear. A single tree on its own is not supported by a community network; it is not sheltered from storms; and its future is genetically isolated. A tree within a forest ecosystem makes connections, bridges generations, leaves a legacy for the future. Now imagine yourself as part of that forest ecosystem. Thin your excess branches, disturb your comforts, form associations, take action for the long-term, and consider the consequences your actions will have in twenty or two hundred years. Imagine a way to live with forests without destroying them.

July 19 —
The pebble edge
of the Willamette River

ed by beavers

POSTSCRIPT

THE INTERSECTION OF
ART AND SCIENCE

It would be possible to describe everything scientifically, but it would make no sense; it would be without meaning, as if you described a Beethoven symphony as a variation of wave pressure.

—ALBERT EINSTEIN, twentieth-century physicist and writer

On this field trip, we imagined a tree falling in the forest and chose to follow it through centuries of decomposition and recomposition into a complex forest ecosystem. Where else might this riddle lead us? Let's take a different path, this time toward the intersection of art and science.

Art and science both begin with wonder. As a verb, "wonder" means to speculate, hypothesize, cogitate, observe. As a noun, "wonder" is related to awe, astonishment, mystery, magic. In both ways, a sense of wonder cultivates the ability to notice details and delight in discovery. Wonder puts art and science on the same adventurous path.

True, the scenery of the temperate rainforest is wonderful to look at, with crystalline streams, volcanic towers, and majestic, moss-draped trees. However, entering the forest with some understanding of how the ecosystem actually lives allows you to become a part of something much bigger than scenery. Knowledge redoubles wonder and prepares you for discovery. Details emerge. Patterns come to light. Suddenly, the scenery materializes into a story

---

A sketchbook is an open door that invites you inside the life of a forest.

with subplots of eco-
system relationships,
characters, drama,
and intrigues.

Science deals
with things that
can be named, mea-
sured, or compared;
art deals with things
that can be seen,
expressed, or re-
membered. Science
tests its interpreta-
tions against natural
laws; art is open to a
viewer's interpreta-
tion. Humans have
a knack for both sci-
ence and art, for be-
ing both rational
and irrational, de-
liberate and spon-
taneous, analyti-
cal and imaginative.
We strive for cer-
tainty but live with
ambiguity. At any
moment of any day,
anywhere, you can

take the measure of your feelings; check the pulse of history; imagine a thing beyond your ability to quantify it. Isn't that wonderful!

A western hemlock takes root atop a decaying Sitka spruce stump in the Siuslaw
National Forest, in a wonderful ongoing pageant of life and death.

Scientific knowledge, as imagined by visualization designer Alberto Cairo, is an island in a sea of mystery. The larger the island of knowledge grows, the longer the shoreline expands where knowledge is washed by the tides of mystery and wonder. An artist and a scientist can be standing at different places along the shoreline, and each expands the island in a slightly different direction. The horizon above the sea of mystery might look different from each perspective, but the view holds the same paradox: the more we learn, the more we see there is to learn. The shoreline lengthens with more wonder.[1]

This recalls Two-Eyed Seeing and the need for humility as we attempt the integration of Indigenous knowledge with Western science. As one eye sees traditional ways of knowing and the other eye sees scientific ways of knowing, together we can see more clearly. Deeper collaborations of these two ways of seeing can change each partner, like the partnership of algae and fungi in becoming a lichen. Something wholly different emerges.

So, what are we seeing? Western science is good at producing a ton of data, but those data need to reflect real landscapes and real inhabitants. We have to go into the forest to see what those pixels really represent. To bring scientists and artists together in the forest is to see creativity at play. Because they come from different disciplines, they don't share the same assumptions or preconceived ideas. Hydrologist Gordon Grant maps how water flows from lava rock; artist Leah Wilson records the color of light reflected off that water. They will expand the way each other sees the world. However, science philosopher Thomas Kuhn emphasizes the *conflict* between two ways of seeing the world, and warns that mutual understanding can happen only if those holding different points of view "can sufficiently refrain from explaining such anomalous behavior as the consequence of mere error or madness."[2] Kuhn wrote the book on scientific revolutions and how scientific traditions are shattered by big, disruptive paradigm shifts. He made it sound so difficult, so uncomfortable. But to see artists and scien-

tists together in the field, questioning observations and reassembling ideas, is playful, pure joy. The community of artists and scientists is a whole greater than its individual parts. It has emergent properties—collaborations, new perspectives, new discoveries—that appear from the heap in unexpected ways to form a creative ecosystem.

The temperate rainforest, alive with natural systems burning, flooding, and blowing up, is the perfect place for creative ecosystems to take root. For geologist Fred Swanson, this dynamic landscape has been a powerful teacher, with old forests and volcanic eruptions throwing down interesting new riddles and "awakening

---

Better than a snapshot, this quick sketch recalls the sounds, smells, and feelings experienced in a moment spent looking at Lookout Creek, in the H. J. Andrews Experimental Forest.

us to 'aha' moments, not as the product of rigorous analysis of long data streams, but really just intuitive leaps of recognition."[3]

As a professor of science communication, I have helped artists add ecological observation to their forest experience, and I have encouraged scientists to imaginatively express their "aha" moments, especially when "aha" becomes "holy shit." My approach is the same with both groups. It starts with drawing what you see. Everyone, you might think, knows how to draw. As little kids, we are fearless artists. But fear creeps in, and too many grown-ups feel panicked when they are asked to "draw what you see." "What do you mean 'what do I see?'" they ask, uncertain how to interpret the image in front of their eyes. It is the same aversion to uncertainty that can lead people to want science to be unambiguous and art to be nice-looking.

Uncertainty is key to both art and science. Uncertainty is a compass that keeps you moving toward greater understanding. It gets you to ask questions, pay attention, make sense. Drawing creates a literal record of what catches your eye, a direct route from seeing to thinking, an excuse to slow down, notice what is around you, and make something of it. By connecting eye and hand in close observation, unexpected details emerge. Artists' sketchbooks and scientists' field notebooks are filled with observations on their way to becoming revelations.

I have had the pleasure of gathering with a group of science-minded artists annually for the last twenty-five years. Like Jerry Franklin's pulses of exploration with the Andrews Forest's scientists, we meet in a different ecosystem each year to sketch, paint, write, and explore together. We look for patterns during the day and share observations at night—individual epiphanies captured in real time. Several of us might sit on the same streambank and draw the same mossy stream in the changing light of the same af-

ternoon. Yet each of our pages will record something different: the reflection of sky on wet boulders, the pull of vine maple toward sunlight, the flash of a kingfisher. No two sketches are the same. No single sketch is enough.

Despite the integration of dynamic landscapes, when I first worked as a fish biologist, field research was segregated by scientific discipline. Fish scientists rarely spoke to forest scientists, much less to climate scientists. We worked in different agencies, measured different things, and spoke different languages. Having been taught that ecology is everything, everywhere, it wasn't always easy to stay focused on fish. I stole moments from my fieldwork to sketch the streams and the logs where I sat to record my measurements. Some of those sketches are in this book. Many of those moments of undisciplined observation have stayed with me.

———————————

This quick sketch caught two old trees in conversation on the Discovery Trail in the H. J. Andrews Experimental Forest.

Now a new creative ecosystem is encouraging communication across lots of perspectives. Young scientists, in particular, are communicating their work like there's no tomorrow: dancing their PhDs; improvising science-based theater; presenting their research in brew pubs with words anyone can understand. If you get stuck in the mud of jargon, mired in words deployed to impress the exalted science priesthood and alienate everyone else, these new science communicators are ready to dig you out of your confusion, throw you a story, a metaphor, or a drawing to grasp as they pull you back to a firm footing of understanding.

Two hundred years ago, von Humboldt recorded his travels with detailed descriptions, illustrated with drawings and maps that greatly expanded the shoreline of wonder. He created a new genre of nature writing that invited readers to imaginatively experience the natural world as if they were exploring alongside him. He wanted the world to see what he saw. An idea that is often attributed to von Humboldt sums up his sense of urgency: the most dangerous worldviews are the worldviews of those who have never viewed the world.

Perhaps the most important benefit of being in the forest is the opportunity to develop an informed worldview by experiencing the world with open eyes. This field trip was meant to help you see the temperate rainforest as a complex ecosystem, crammed with structures and functions that sustain it through tumultuous events over a very long time. From the messy tangle of a forest's features emerge endless wonders, much more than the sum of its parts. Open your eyes to them. As our world gets hotter and more crowded, it is more important than ever to see a wild forest as an essential part of life on Earth rather than as an untapped resource to feed insatiable wealth and growth. We must not cash in our legacy of wild forests. It matters what we leave behind.

NOTES

1 You Are Here

Epigraph: David Douglas cited in Jack Nisbet, *David Douglas: A Naturalist at Work: An Illustrated Exploration across Two Centuries in the Pacific Northwest* (Seattle: Sasquatch Books, 2012), 101.

1. Beverly Law, "Wildfire in a Warming World: Opportunities to Improve Community Collaboration, Climate Resilience, and Workforce Capacity" (statement to the U.S. House of Representatives Subcommittee on National Parks, Forests and Public Lands, Washington, DC, April 29, 2021).

2. J. F. Franklin, "Vegetation of the Douglas-Fir Region," chap. 4 in *Forest Soils of the Douglas-Fir Region* (Pullman: Washington State University Cooperative Extension, 1979).

3. It's tricky to identify the traditional names of volcanic peaks because different tribes had different names, sometimes multiple names, for the multiple personalities of each mountain.

4. J. F. Franklin and C. T. Dyrness, *Natural Vegetation of Oregon and Washington* (Corvallis: Oregon State University Press, 1973, plus 1988 supp.); D. A. DellaSala et al., "Temperate and Boreal Rainforests of the Pacific Coast of North America," in *Temperate and Boreal Rainforests of the World* (Washington, DC: Island Press, 2010), 42–81.

5. Charles L. Bolsinger and Karen L. Waddell, *Area of Old-Growth Forests in Forest Service California, Oregon, and Washington* (Portland: USDA Forest Service Pacific Northwest Research Station, 1993).

6. The White House, "Executive Order on Strengthening the Nation's Forests, Communities, and Local Economies," April 22, 2022, https://www.whitehouse.gov/briefing-room.

2 You Enter a Land of Big, Old Trees

Epigraph: Jerry Franklin, email message to author, January 10, 2022. Jerry Franklin is a professor emeritus of forest ecology at the University of Washington and a longtime research lead at H. J. Andrews Experimental Forest.

1. Franklin, email message to author, January 10, 2022.

2. M. L. Herring, "A Community of Equals: Diversity in the Old-Growth Forest," *Pacific Discovery* 44, no. 4 (Fall 1991): 10.

3. R. H. Waring and J. F. Franklin, "Evergreen Coniferous Forests of the Pacific Northwest," *Science* 204, no. 4400 (June 1979): 1381.

4. Fabien L. Condamine et al., "The Rise of Angiosperms Pushed Conifers to Decline during Global Cooling," *PNAS* 117, no. 46 (November 17, 2020): 28867–75.

5. S. C. Sillett et al., "Development and Dominance of Douglas-Fir in North American Rainforests," *Forest Ecology and Management* 429 (2018): 93–114.

6. M. L. Herring and S. E. Greene, *Forest of Time* (Corvallis: Oregon State University Press 2007), 63.

7. Roy R. Silen and Leonard R. Woike, *The Wind River Arboretum,* USDA Forest Service Pacific Northwest Research Station Paper no. 33 (Portland: Pacific Northwest Forest and Range Experiment Station, 1959); J. V. Hoffmann, "Adaptation in Douglas Fir," *Ecology* 2, no. 2 (1921): 131; John Kirkland and Brad St. Clair, "The 1912 Douglas-Fir Heredity Study: Lessons from a Century of Experience" *Science Findings* 235 (2021): 5.

8. Barbara Lachenbruch, interview with the author, September 3, 2021; Barbara Lachenbruch, email message to author, November 10, 2023.

9. Jean-Christophe Domec et al., "Maximum Height in a Conifer Is Associated with Conflicting Requirements for Xylem Design," *PNAS* 105, no. 33 (August 19, 2008): 12069–74.

10. "Journals Kept by David Douglas during His Travels in North America 1823–1827," Royal Horticultural Society, Internet Archive, https://archive.org.

3 You Stand on a Rollicking Foundation

Epigraph: Gordon Grant, "Geology Is Destiny: Cold Waters Run Deep," *Science Findings* 49 (December 2002): 1.

1. Ray Wells et al., "Geologic History of Siletzia," *Geosphere* 10, no. 4 (2014): 692–719.

2. Kathryn Schulz, "The Really Big One," *New Yorker,* July 20, 2015, 52–59.

3. National Oceanic and Atmospheric Administration, "JetStream Max: Cascadia Subduction Zone," https://www.noaa.gov.

4. Chris Goldfinger, "Toast, Tsunamis and the Really Big One" (TEDx Portland, July 5, 2016).

5. Patricia Whereat Phillips, "Tsunamis and Floods in Coos Bay Mythology," *Oregon Historical Quarterly* 108, no. 2 (Summer 2007): 190–92.

6. Oregon State University, "Odds Are About 1-in-3 That Mega-Earthquake Will Hit Pacific Northwest in Next 50 Years, Scientists Say," ScienceDaily, May 25, 2010, https://www.sciencedaily.com.

7. Christina Tague and Gordon E. Grant, "A Geological Framework for Interpreting the Low-Flow Regimes of Cascade Streams, Willamette River Basin, Or-

egon," *Water Resources Research* 40, no. 4 (April 2004), https://doi.org/10.1029
/2003WR002629.

8. Cascades Volcano Observatory, "Columbia River Basalt Group Stretches
from Oregon to Idaho," U.S. Geological Survey, December 7, 2023, https://www
.usgs.gov/observatories.

9. National Ecological Observatory Network, "Wind River Experimental Forest NEON (WREF) Soil Descriptions," March 2019, https://www.neonscience.org.

10. Fred Swanson, "How Do We Best Study Rare Disturbances?" (Starker
Lecture Series panel discussion, College of Forestry, Oregon State University,
March 10, 2021); Fred Swanson, oral history, November 15, 2013, Oregon State Uni-
versity Archives, 00:02.00–00:03:00.

11. Frederick J. Swanson, "Languages of Volcanic Landscapes," in *In the Blast
Zone: Catastrophe and Renewal on Mount St. Helens,* ed. Charles Goodrich, Kath-
leen Dean Moore, and Frederick Swanson (Corvallis: Oregon State University
Press, 2008), 105.

12. Jerry F. Franklin, "Evolutionary Impacts of a Blasted Landscape," in Goodrich et al., *In the Blast Zone,* 63.

13. Charlie Crisafulli, "Volcano Ecology: Flourishing on the Flanks of Mount
St. Helens," *Science Findings* 190 (October 2016): 3. Crisafulli, a retired research
ecologist at the USDA Pacific Northwest Research Station, spent his entire career
studying the renewal of Mount St. Helens.

14. Warren Cornwall, "Mount St. Helens 40 Years Later: What We've
Learned, and Still Don't Know," *Science,* May 18, 2020, https://www.science.org.

4 You Follow Steps toward a Very Long Life

Epigraph: John Muir, "The Forests of Oregon and Their Inhabitants" (1888),
repr. by the Sierra Club, "John Muir Exhibit," https://vault.sierraclub.org/john
_muir_exhibit.

1. Eugene P. Odum, *Fundamentals of Ecology,* 3rd ed. (Philadelphia: W. B.
Saunders, 1972), 251.

2. Jerry Franklin et al., "Forest Dynamics," in *Ecological Forest Management*
(Long Grove, IL: Waveland Press, 2018), 55–64.

3. Jerry F. Franklin, interview with author, February 18, 2022; Jerry F. Frank-
lin, "Ecological Attributes of Mature Douglas-Fir–Western Hemlock Forests: The
Majestic and Transformative Mature Forests of Douglas-Fir!," unpublished draft 2,
January 7, 2022.

4. Franklin et al., "Forest Dynamics."

5. Muir, "Forests of Oregon."

6. William G. Robbins, *Landscapes of Conflict: The Oregon Story, 1940–2000*
(Seattle: University of Washington Press, 2004), 194–204.

7. Carl Segerstrom, "Can a Campaign for Nature and Community Rights Stop
Aerial Spraying in Oregon?," *High Country News,* October 23, 2019, https://www

.hcn.org; Laurie Bernstein et al., "Oregon's Industrial Forests and Herbicide Use: A Case Study of Risk to People, Drinking Water and Salmon," Beyond Toxics, December 2013, https://www.beyondtoxics.org.

8. Robert Leo Hielman, "With a Human Face: When Hoedads Walked the Earth," *Oregon Quarterly* 91 (Autumn 2011): 36–43.

9. Andy Kerr, "Oregon and Washington Raw Log Exports: Exporting Jobs and a Subsidy to Domestic Mills," Larch Company Occasional Paper no. 10, February 2012, https://www.andykerr.net/larch-occasional-papers.

10. Engineering Channel, "How Feller Buncher Works," YouTube, https://www.youtube.com.

11. J. E. M. Watson et al., "The Exceptional Value of Intact Forest Ecosystems," *Nature Ecology and Evolution* 2 (2018): 599–610.

5 You Glimpse What It Means to Be Old

Epigraph: Kathleen Dean Moore and Michael Paul Nelson, "The Perilous and Important Art of Definition: The Case of the Old-Growth Forest," *Frontiers in Ecology and the Environment* 21, no. 6 (August 2023): 264–65.

1. Rachel Carson, *Silent Spring* (New York: Houghton Mifflin, 1962); Paul and Anne Ehrlich, *The Population Bomb* (New York: Sierra Club/Ballantine, 1968); Garrett Hardin, "The Tragedy of the Commons," *Science* 162, no. 3859 (December 13, 1968): 1243–48.

2. Meir Rinde, "Richard Nixon and the Rise of American Environmentalism," Science History Institute, June 2, 2017, https://www.sciencehistory.org.

3. William G. Robbins, *Landscapes of Conflict: The Oregon Story, 1940–2000* (Seattle: University of Washington Press 2004), 205–6.

4. Mark E. Harmon and David M. Bell, "Mortality in Forested Ecosystems: Suggested Conceptual Advances," *Forests* 11, no. 5 (May 2020): 572.

5. Matt Betts, interview with the author, February 14, 2023; Matthew G. Betts et al., "Forest Degradation Drives Widespread Avian Habitat and Population Declines," *Nature Ecology and Evolution* 6 (April 28, 2022): 709–19; Hankyu Kim et al., "Forest Microclimate and Composition Mediate Long-Term Trends of Breeding Bird Populations," *Global Change Biology* 28, no. 21 (September 6, 2022): 6180–93.

6. John Muir, "The Douglas Squirrel," in *The Mountains of California* (New York: Century, 1894), repr. by the Sierra Club, "The John Muir Exhibit," https://vault.sierraclub.org/john_muir_exhibit/.

7. Mark E. Harmon et al., "Ecology of Coarse Woody Debris in Temperate Ecosystems," *Advances in Ecological Research* 15 (December 1986): 133–276.

8. Glenn A. Christiansen et al., *Oregon Forest Ecosystems Carbon Inventory: 2001–2016* (Portland: USDA Forest Service Pacific Northwest Research Station, 2019); Jerry Melillo, "Forests and Climate Change," Climate Portal (Massachusetts Institute of Technology), https://climate.mit.edu.

9. Stephen C. Sillett et al., "Comparative Development of the Four Tallest Co-nifer Species," *Forest Ecology and Management* 480 (January 2021): 1186–88; Ste-phen C. Sillett, "Development and Dominance of Douglas-Fir in North American Rainforests," *Forest Ecology and Management* 429 (2018): 93–114.

10. Steve Sillett, interview with the author, February 12, 2014.

11. Rocky Mountain Tree-Ring Research, "Old List: A Database of Old Trees," https://www.rmtrr.org; Craig Welch, "What's the Oldest Tree on Earth—and Will It Survive Climate Change?," *National Geographic,* May 2022, https://www.nationalgeographic.com.

12. Although people were outraged that the ancient tree was killed, the re-searcher's mistake revealed the unimaginable longevity possible in trees and helped to establish Great Basin National Park to protect the bristlecones.

13. Bruce Marcot et al., *Wildlife-Habitat Relationships: Concepts and Applica-tions* (Madison: University of Wisconsin Press, 1998).

14. Eric Forsman, oral history, December 5, 2016, Oregon State University Ar-chives, 1:50:00–1:60:00 and 00.40.00–00.53.00.

15. Forsman, oral history.

16. Forsman, oral history.

17. Sean Nealon, "Removal of Barred Owls Slows Decline of Iconic Spot-ted Owls in Pacific Northwest, Study Finds," Oregon State University newsroom, July 19, 2021, https://today.oregonstate.edu/news.

6 You Explore the Forest Underworld

Epigraph: Mark Harmon, email message to author, December 14, 2021. Mark Harmon is a professor emeritus of forest ecosystems at Oregon State University.

1. Chris Maser et al., *The Seen and Unseen World of the Fallen Tree* (Portland: USDA Forest Service Pacific Northwest Research Station, 1984).

2. Mark E. Harmon et al., "Carbon Concentration of Standing and Downed Woody Detritus: Effects of Tree Taxa, Decay Class, Position, and Tissue Type," *Forest Ecology and Management* 291, no. 1 (March 2013), https://doi.org/10.1016/j.foreco.2012.11.046.

3. Cheryl Dybas, "Morticulture: Forests of the Living Dead (an interview with Mark Harmon)," U.S. National Science Foundation, October 19, 2015, https://new.nsf.gov/news.

4. Jerry Franklin, email message to author, January 25, 2022.

5. Suzanne W. Simard et al., "Net Transfer of Carbon between Ectomycorrhi-zal Tree Species in the Field," *Nature,* August 7, 1997, 579–82. Suzanne Simard's work launched a bioregional study in British Columbia, the Mother Tree Project, to identify sustainable harvesting and regeneration practices that minimize nega-tive effects on forest carbon, biodiversity, and other ecological processes.

6. Simard et al., "Net Transfer of Carbon"; Merlin Sheldrake, *Entangled Life:*

How Fungi Make Our World, Change Our Minds, and Shape Our Futures (New York: Random House, 2020).

7. John Kirkland et al., "The More the Mossier: Using Community Science to Map Air Quality in Environmental Justice Investigations," *Science Findings* 255 (March 2023): 4.

8. Rachel Cooke, "Robin Wall Kimmerer: 'Mosses are a model of how we might live,'" *Guardian,* June 19, 2021.

9. Robin W. Kimmerer, *Gathering Moss: A Natural and Cultural History of Mosses* (Corvallis: Oregon State University Press, 2003), 150.

7 You Drift to the Top of the Canopy

Epigraph: Nalini Nadkarni, email message to author, October 2023. Nalini Nadkarni is a pioneering tree canopy researcher and retired professor of forest ecology at the University of Utah.

1. Robert Jimison, "Why We All Need Green in Our Lives," CNN, June 5, 2017, https://www.cnn.com; listen to physicist Richard Feynman explain why trees are made from air at ScienceChannel9000, "Richard Feynman: Biology of a Tree," YouTube, https://www.youtube.com; Michael Shermer, "Unweaving the Heart," *Scientific American,* October 2005, 35–36.

2. Qing Li, "Effect of Forest Bathing Trips on Human Immune Function," *Environmental Health and Preventive Medicine* 15, no. 1 (January 2010): 9–17; Jo Barton and Jules Pretty, "What Is the Best Dose of Nature and Green Exercise for Improving Mental Health? A Multi-Study Analysis," *Environmental Science and Technology* 44 (May 2010): 3947–55.

3. Damon B. Lesmeister and Julianna M. A. Jenkins, "Integrating New Technologies to Broaden the Scope of Northern Spotted Owl Monitoring and Linkage with USDA Forest Inventory Data," *Frontiers in Global Change* 5 (October 2022): 966–78; Christopher Mims, "Alexa for Animals: AI Is Teaching Us How Creatures Communicate," *Wall Street Journal,* March 19, 2022.

4. Bruce McCune, "Gradients in Epiphyte Biomass in Three Pseudotsuga-Tsuga Forests of Different Ages in Western Oregon and Washington," *Bryologist* 96, no. 3 (Autumn 1993): 405–11.

5. George Carroll, "Fungal Endophytes in Stems and Leaves: From Latent Pathogen to Mutualistic Symbiont," *Ecology* 69, no. 1 (February 1988): 2–9; J. Mishra, E. Khare, and N. K. Arora, "Multifaceted Interactions between Endophytes and Plant: Developments and Prospects," *Frontiers of Microbiology* 9 (November 15, 2018), https://doi.org/10.3389/fmicb.2018.02732.

6. Nicholas J. Talbot, "Plant Immunity: A Little Help from Fungal Friends," *Current Biology* 25, no. 22 (November 16, 2015): 1074–76.

7. Australian National Herbarium, "Beatrix Potter," https://www.anbg.gov.au.

8. David C. Shaw et al., "The Vertical Occurrence of Small Birds in an Old-Growth Douglas-Fir-Western Hemlock Forest Stand," *Northwest Science* 76, no. 4 (May 28, 2002): 322–34.

9. Sarah J. K. Frey et al., "Spatial Models Reveal the Microclimatic Buffering Capacity of Old-Growth Forests," *Science Advances* 2, no. 4 (2016), https://doi.org/10.1126/sciadv.1501392.

10. Frey et al., "Spatial Models."

11. See a photograph of Diane Nielson scaling a tall Douglas-fir: Oregon State University Digital Archives, "Diane Nielson Climbing Tree, Oregon State University, Corvallis, May 1971," https://oregondigital.org/catalog/oregondigital:9z903g47g.

12. Nalini Nadkarni, *Between Earth and Sky: Our Intimate Connection to Trees* (Berkeley: University of California Press, 2008).

13. David C. Shaw and Sarah E. Greene, "Wind River Canopy Crane Research Facility and Wind River Experimental Forest," *Bulletin of the Ecological Society of America* 84, no. 3 (July 2003): 115–21.

14. NEON Observatory blog, "Measuring the Impact of Nature-Based Climate Solutions with Flux Towers," April 20, 2022, https://www.neonscience.org.

15. Julia A. Jones, "Hydrologic Responses to Climate Change: Considering Geographic Context and Alternative Hypotheses," *Hydrologic Processes* 25, no. 12 (2011), https://doi.org/10.1002/hyp.8004; Timothy D. Perry and Julia A. Jones, "Summer Streamflow Deficits from Regenerating Douglas-Fir Forest in the Pacific Northwest, USA," *Ecohydrology* 10, no. 2 (September 5, 2016), https://doi.org/10.1002/eco.1790.

16. Julia Jones, interview with the author, October 15, 2019.

8 You Float Down the Watershed

Epigraph: Sherri Johnson, interview with author, August 12, 2022. Sherri Johnson is a research ecologist with the USDA Forest Service Pacific Northwest Research Station, and was the U.S. Forest Service lead scientist at the H. J. Andrews Experimental Forest for more than twenty years.

1. Stephen Dow Beckham, "Changes in the Land, Oregon 1800 to 2020" (Champinefu Lecture Series, Corvallis, Oregon, October 14, 2020); Rebecca R. Miller, "Is the Past Present? Historical Splash-Dam Mapping and Stream Disturbance Detection in the Oregon Coastal Province" (master's thesis, Oregon State University, September 23, 2010), 14.

2. Thomas E. Nickelson et al., "Seasonal Changes in Habitat Use by Juvenile Coho Salmon in Oregon Coastal Streams," *Canadian Journal of Fisheries and Aquatic Sciences* 49, no. 2 (April 1992): 783–89.

3. Ben Goldfarb, *Eager: The Surprising Secret Life of Beavers and Why They Matter* (White River Junction, VT: Chelsea Green Publishing, 2019); KUOW, "Parachuting Beavers" (1948), YouTube, https://www.youtube.com; Julia Zorthian, "The True History behind Idaho's Parachuting Beavers," *Time,* October 23, 2015, https://time.com.

4. Johnson, interview with author, August 12, 2022.

5. Thomas P. Quinn et al., "A Multidecade Experiment Shows That Fertiliza-

tion by Salmon Carcasses Enhanced Tree Growth in the Riparian Zone," *Ecology* 99, no. 11 (October 23, 2018): 2433–41; Thomas E. Reimchen and Estelle Arbellay, "Influence of Spawning Salmon on Tree-Ring Width, Isotopic Nitrogen, and Total Nitrogen in Old-Growth Sitka Spruce from Coastal British Columbia," *Canadian Journal of Forest Research* 49, no. 9 (July 2019): 1078–86.

6. Chris Maser and James R. Sedell, *From the Forest to the Sea: The Ecology of Wood in Streams, Rivers, Estuaries and Oceans* (Delray Beach, FL: St. Lucie Press, 1994); Ed Yong, "The Marine Creatures That Only Live on Land Plants," *National Geographic,* April 8, 2014, https://www.nationalgeographic.com.

7. Brooke Penaluna, interview with the author, October 30, 2022; Brooke E. Penaluna et al., "Better Boundaries: Identifying the Upper Extent of Fish Distributions in Forested Streams Using eDNA and Electrofishing," *Freshwater Ecology* (January 20, 2021), https://doi.org/10.1002/ecs2.3332. Brooke Penaluna is a research fisheries biologist with the USDA Pacific Northwest Research Station and the lead scientist at H. J. Andrews Experimental Forest.

8. Jim Lichatowich, *Salmon without Rivers: A History of the Pacific Salmon Crisis* (Washington, DC: Island Press, 1999); Cheryl Lynn Dybas, "Ode to a Codfish," *BioScience* 56, no. 3 (March 2006): 181–84; Chad C. Meengs and Robert T. Lackey, "Estimating the Size of Historical Oregon Salmon Runs," *Reviews in Fisheries Science* 13, no. 1 (2005): 51–66.

9. Willa Nehlsen et al., "Pacific Salmon at the Crossroads: Stocks at Risk from California, Oregon, Idaho, and Washington," *Fisheries* 16, no. 2 (1991): 4–21.

10. Hongxia Zhang et al., "Seawater Exposure Causes Hydraulic Damage in Dying Sitka-Spruce Trees," *Plant Physiology* 187, no. 2 (October 2021): 873–85.

11. Jane Braxton Little, "Notes from the (Water) Underground," *High Country News*, January 17, 2009, https://www.hcn.org; USDA Forest Service, "Meet the Forest Service," https://www.fs.usda.gov.

12. Marie Oliver, "Liberated Rivers: Lessons From 40 Years of Dam Removal," *Science Findings* 193 (February 2017): 1–5.

13. Bellamy Pailthorp, "For First Time since Dam Removal, a Fishery Opens on the Elwha," KNKX Public Radio, October 11, 2023.

9 You Witness Disturbing Events

Epigraph: Fred Swanson, oral history, Oregon State University Archives, September 6, 1996, 00.48.00.

1. Jerry Franklin et al., "Forest Dynamics," in *Ecological Forest Management* (Long Grove, IL: Waveland Press, 2018), 48–89.

2. Matthew J. Reilly et al., "Cascadia Burning: The Historic, but Not Historically Unprecedented, 2020 Wildfires in the Pacific Northwest, USA," *Ecosphere* 13, no. 6 (June 2022): 6383–403.

3. James D. Johnston et al., "Exceptional Variability in Historical Fire Regimes across a Western Cascades Landscape, Oregon, USA," *Ecosphere* 14, no. 12 (December 20, 2023), https://doi.org/10.1002/ecs2.4735; Alan J. Tepley et al., "Fire-

Mediated Pathways of Stand Development in Douglas-Fir/Western Hemlock Forests of the Pacific Northwest, USA," *Ecology* 9, no. 8 (August 2013): 1729–43.

4. Jeffrey E. Stern, "The Battle of High Hill," *Atlantic,* August 30, 2021, https://www.theatlantic.com/magazine.

5. National Oceanic and Atmospheric Administration, "Fire-Breathing Storm Systems," October 19, 2010, https://earthobservatory.nasa.gov.

6. Reilly et al., "Cascadia Burning"; also, for a thorough study of western fire ecology, see James K. Agee, *Fire Ecology of Pacific Northwest Forests* (Washington, DC: Island Press, 2013).

7. T. T. Munger, "Out of the Ashes of the Nestucca," *American Forests* 40 (1944): 366–68.

8. USDA Forest Service Northern Research Station, "Understanding the Wildland-Urban Interface (1990–2020)," USDA Forest Service Northern Research Station, September 20, 2023, https://storymaps.arcgis.com.

9. Daniel C. Donato et al., "Corralling a Black Swan: Natural Range of Variation in a Forest Landscape Driven by Rare, Extreme Events," *Ecological Applications* 30, no. 1 (October 2019), https://doi.org/10.1002/eap.2013.

10. Harold S. J. Zald and Christopher J. Dunn, "Severe Fire Weather and Intensive Forest Management Increase Fire Severity in a Multi-Ownership Landscape," *Ecological Applications* 28 (April 26, 2018): 1068–80.

11. Alai Reyes-Santos and Joe Scott, "Indigenous Fire Practices Can Help Oregon Wildfires, Land Management," *Register-Guard,* March 17, 2022, https://www.registerguard.com.

12. Timothy Ingalsbee, "Greenfire Revolution: The Ancient/Future Paradigm of Ecological Fire Management" (presentation at Spring Creek Project, Oregon State University, February 11, 2022); Timothy Ingalsbee, "Whither the Paradigm Shift? Large Wildland Fires and Wildlife Paradox Offer Opportunities for a New Paradigm of Ecological Fire Management," *International Journal of Wildland Fire* 26 (July 2017): 557–61.

13. . . . or from floods, volcanic eruptions, or windstorms, but you get the point.

14. Will Shenton, "As Climate Warms, Overlapping Wildfires Are Changing Forest Resilience," University of Washington College of the Environment, April 5, 2023, https://environment.uw.edu.

15. John Dodge, *A Deadly Wind: The 1962 Columbus Day Storm* (Corvallis: Oregon State University Press, 2018).

16. Jeff LaLande, "The Columbus Day Storm (1962)," The Oregon Encyclopedia, https://www.oregonencyclopedia.org.

17. Western Regional Climate Center, "Oregon's Top 10 Weather Events of the 1900s," https://wrcc.dri.edu.

18. Alan Honick, *Torrents of Change* (video), https://alanhonick.com.

19. Jonathan Thompson and Gordon Grant, "Does Wood Slow Down 'Sludge Dragons'?," *Science Findings* 86 (September 2006), https://www.fs.usda.gov/research/pnw.

20. Ken Armstrong et al., "'Unforeseen' Risk of Slide? Warnings Go Back De-

cades," *Seattle Times,* March 25, 2014; Timothy Egan, "A Mudslide, Foretold," *New York Times,* March 29, 2014.

21. Jerry Franklin, email message to author, January 25, 2022.

22. Franklin, email message to author, January 25, 2022.

23. Robert Sullivan, foreword in *Hemlock: A Forest Giant on the Edge,* ed. David R. Foster et al. (New Haven: Yale University Press, 2014), xix.

10 You Find Human Fingerprints

Epigraph: K. Norman Johnson (remarks during a presentation of "The Making of the Northwest Plan," Oregon State University College of Forestry, May 10, 2023).

1. Bill Clinton, "Remarks Concluding the First Roundtable Discussion of the Forest Conference in Portland, April 2, 1993," in *Public Papers of the Presidents of the United States, William J. Clinton* (Washington, DC: U.S. Government Printing Office, 1994–), book 1, 386–87.

2. William G. Robbins, "The First Peoples," Oregon History Project, 2002, https://www.oregonhistoryproject.org; Kenneth Greg Watson, "Native Americans of Puget Sound—A Brief History of the First People and Their Cultures," History-Link, June 29, 1999, https://www.historylink.org; Alexander Koch et al., "European Colonization of the Americas Killed 10 Percent of World Population and Caused Global Cooling," The World, January 31, 2019, https://theworld.org.

3. Seth Zuckerman and Edward C. Wolf, *Salmon Nation: People, Fish, and Our Common Home* (Corvallis: Oregon State University Press, 2003); Jeri Chase, "Western Redcedar, Tree of Life," *Forests for Oregon* (Fall 2008): 18–19.

4. Union of British Columbia Indian Chiefs, "Background on Indian Reserves in British Columbia," https://ourhomesarebleeding.ubcic.bc.ca.

5. Stephen F. Arno and Carl E. Fiedler, *Douglas Fir: The Story of the West's Most Remarkable Tree* (Seattle: Mountaineers Books, 2020), 73, 76.

6. William G. Robbins, *Landscapes of Promise: The Oregon Story, 1800–1940* (Seattle: University of Washington Press 1997), 226–28.

7. National Park Service, "The Spruce Production Division: Fort Vancouver National Historic Site," https://www.nps.gov/articles.

8. Margaret Herring and Sarah Greene, *Forest of Time* (Corvallis: Oregon State University Press, 2007), 63–64.

9. William G. Robbins, *A Place for Inquiry, A Place for Wonder* (Corvallis: Oregon State University Press, 2020), 17, 25–27. This is the definitive history of the H. J. Andrews Experimental Forest.

10. National Academy of Sciences, "The International Biological Program (IBP), 1964–1974," https://www.nasonline.org.

11. Max G. Geier, *Necessary Work: Discovering Old Forests, New Outlooks, and Community on the Andrews Forest* (Portland: USDA Forest Service Pacific Northwest Research Station, 2007); Jon R. Luoma, *The Hidden Forest: The Biography of an Ecosystem,* 2nd ed. (Corvallis: Oregon State University Press, 2006).

12. James Sedell, "Science Tribes of Mt. St. Helens," in *In the Blast Zone,* ed. Charles Goodrich, Kathleen Dean Moore, and Frederick Swanson (Corvallis: Oregon State University Press, 2008), 86.

13. Herring and Greene, *Forest of Time,* 119–22. Defining "old growth" was eventually published as Jerry Franklin et al., *Ecological Characteristics of Old-Growth Forests* (Portland: USDA Forest Service Pacific Northwest Research Station, 1981).

14. National Science Foundation, LTER Network, https://lternet.edu.

15. Mark E. Harmon, "200-Year Decomposition Study at Andrews Forest," *LTER Network News* 18, no. 2 (2005), https://lternet.edu.

16. Jerry F. Franklin, foreword to *Bioregional Assessments: Science at the Crossroads of Management and Policy,* ed. K. Norman Johnson et al. (Washington, DC: Island Press, 1999), xi–xiii.

17. Eric D. Forsman et al., "Distribution and Biology of the Spotted Owl in Oregon," *Wildlife Monographs* 87 (April 1984): 3–64; Ted Gup, "Who Gives a Hoot?," *Time,* June 25, 1990, 56.

18. Gup, "Who Gives a Hoot?"

19. Andy Kerr, "Starting the Fight and Finishing the Job," in *Old Growth in a New World,* ed. Thomas A. Spies and Sally L. Duncan (Washington, DC: Island Press, 2009), 131–32. Andy Kerr is an environmental activist in Oregon.

20. K. Norman Johnson et al., *The Making of the Northwest Forest Plan: The Wild Science of Saving Old-Growth Ecosystems* (Corvallis: Oregon State University Press, 2023), 116–20. This thorough history is an insider's view of the tumultuous, litigious events from the owl wars to the Northwest Forest Plan.

21. Willa Nehlsen et al., "Pacific Salmon at the Crossroads: Stocks at Risk from California, Oregon, Idaho, and Washington," *Fisheries* 16, no. 2 (March–April 1991): 4–21.

22. Johnson et al., *Making of the Northwest Forest Plan,* 125.

23. Johnson et al., *Making of the Northwest Forest Plan,* 170.

24. Norm Johnson, oral history, Oregon State University Archives, November 29, 2016, 01:27:00.

25. Jack Ward Thomas, "Learning from the Past and Moving to the Future," in Johnson, *Bioregional Assessments,* 11–25.

26. Jerry Franklin (remarks during a presentation of "The Making of the Northwest Plan," Oregon State University College of Forestry, May 10, 2023).

27. Headwaters Economics, "The Transition from Western Timber Dependence: Lessons for Counties," December 2017, headwaterseconomics.org.

28. Clinton, "Remarks," 386–87.

29. Johnson et al., *Making of the Northwest Forest Plan,* 253–58.

30. Tony Davis, "Last Line of Defense," *High Country News,* September 2, 1996, https://www.hcn.org.

31. James Furnish, *Toward a Natural Forest* (Corvallis: Oregon State University Press, 2015); Furnish, interview with the author, October 14, 2023.

32. Thomas A. Spies et al., "Twenty-Five Years of the Northwest Forest Plan:

What Have We Learned?," *Frontiers in Ecology and the Environment* 35 (August 28, 2019): 1319.

33. Environmental and Energy Law Program (Harvard University), "Regulatory Tracker: Alaska Roadless Rule," https://eelp.law.harvard.edu.

34. Furnish, interview with the author, October 14, 2023.

35. Johnson et al., *Making of the Northwest Forest Plan,* 30–31.

36. This statement was signed by Secretary of Agriculture James Wilson on February 1, 1905. It is addressed to "The Forester," the chief of the newly created Forest Service, Gifford Pinchot, and it is generally assumed that Pinchot himself wrote the letter. Forest History Society, "Wilson Letter," https://foresthistory.org.

37. Cristina Eisenberg, interview with the author, February 12, 2024; Sheraz Sadiq, "OSU College of Forestry Partners with Oregon Tribes to Restore Lands and Support Underrepresented Students," Oregon Public Broadcasting, September 13, 2022, https://www.opb.org.

11 You Balance on a Rapidly Changing Planet

Epigraph: Beverly E. Law et al., "Creating Strategic Reserves to Protect Forest Carbon and Reduce Biodiversity Losses in the United States," *Land* 11, no. 5 (May 11, 2022): 721.

1. NOAA Global Monitoring Laboratory, "Trends in Atmospheric Carbon Dioxide," https://gml.noaa.gov.

2. Christopher Flavelle et al., "Climate Shocks Are Making Parts of America Uninsurable," *New York Times,* May 31, 2023.

3. Elizabeth Kolbert, *Under a White Sky: The Nature of the Future* (New York: Crown, 2021).

4. Meilan Solly, "The U.S. Loses a Football Field-Sized Patch of Nature Every 30 Seconds," *Smithsonian,* August 12, 2019, https://www.smithsonianmag.com.

5. Mikaela Weisse and Elizabeth Goldman, "Primary Rainforest Destruction Increased 12% from 2019 to 2020," *Global Forest Watch,* March 31, 2021, https://www.globalforestwatch.org.

6. Roger Worthington, "I Pledged $1 Million to Plant New Trees. I Wish I Could Invest the Money in Saving Old Ones," *New York Times,* June 14, 2023.

7. Brooke Hirsheimer, "69% Average Decline in Wildlife Populations since 1970, Says New WWF Report," press release, World Wildlife Fund, October 13, 2022, https://www.worldwildlife.org.

8. Kim Nelson, interview with the author, August 30, 2023; Juliet Grable, "From Sea to Tree, Scientists Are Tracking Marbled Murrelets with Rising Precision," *Audubon,* Fall 2018, https://www.audubon.org/magazine.

9. UN Environment Programme, "COP15 Ends with Landmark Biodiversity Agreement," December 20, 2022, https://www.unep.org.

10. Office of the Director of National Intelligence, "Global Water Security," Intelligence Community Assessment, ICA 2012-08, February 2, 2012, https://

www.dni.gov/files/documents/Special%20Report_ICA%20Global%20Water%20Security.pdf.

11. Emmanuel Raju et al., "Stop Blaming the Climate for Disasters," *Communications Earth and Environment* 3, no. 1 (January 10, 2022), https://doi.org/10.1038/s43247-021-00332-2.

12. Fred Pearce, "It's Not Just Climate: Are We Ignoring Other Causes of Disasters?," Yale Environment 360, February 8, 2022, https://e360.yale.edu.

13. Thomas A. Spies et al., "Climate Change Adaptation Strategies for Federal Forests of the Pacific Northwest, USA: Ecological, Policy, and Socio-Economic Perspectives," *Landscape Ecology* 25 (May 6, 2010): 1185–99.

14. Jerry F. Franklin et al., "Effects of Global Climate Change on Forests in Northwestern North America," in *Forests in Northwestern North America* (New Haven: Yale University Press, 1992), 224–60.

15. Andy Kerr, "Smoke Happens," *Andy Kerr's Public Lands Blog*, October 10, 2018, https://www.andykerr.net.

16. Joshua Halofsky et al., "The Nature of the Beast: Examining Climate Adaptation Options in Forests with Stand-Replacing Fire Regimes," *Ecosphere* 9, no. 3 (March 2018), https://doi.org/10.1002/ecs2.2140.

17. J. R. R. Tolkien, *The Two Towers* (New York: Ballentine Books, 1965), 223.

18. The White House, "Fact Sheet: President Biden Signs Executive Order to Strengthen America's Forests, Boost Wildfire Resilience, and Combat Global Deforestation," April 22, 2022, https://www.whitehouse.gov/briefing-room.

19. Matthew J. Reilly et al., "Cascadia Burning: The Historic, but Not Historically Unprecedented, 2020 Wildfires in the Pacific Northwest, USA," *Ecosphere* 13, no. 6 (June 2022): 6383–403; Halofsky et al., "Nature of the Beast"; Daniel C. Donato et al., "Corralling a Black Swan: Natural Range of Variation in a Forest Landscape Driven by Rare, Extreme Events," *Ecological Applications* 30, no. 1 (October 2019), https://doi.org/10.1002/eap.2013.

20. L. V. Gatti et al., "Amazonia as a Carbon Source Linked to Deforestation and Climate Change," *Nature,* July 14, 2021, 388–93.

21. Task Force on National Greenhouse Gas Inventories, "2006 IPCC Guidelines for National Greenhouse Gas Inventories," IPCC, 2.17, Table 2.2, https://www.ipcc-nggip.iges.or.jp.

22. Mira Rojanasakul, "Wildfire Smoke Is Erasing Progress on Clean Air," *New York Times,* September 22, 2022; Climate Central, "Wildfire Smoke: Nationwide Health Risk," October 3, 2023, https://www.climatecentral.org.

23. Sharon Guynup, "COP26: Surging Wood Pellet Industry Threatens Climate, Say Experts," Mongabay, November 2021, https://news.mongabay.com.

24. Beverly Law, "Wildfire in a Warming World: Opportunities to Improve Community Collaboration, Climate Resilience, and Workforce Capacity" (statement to the U.S. House of Representatives Subcommittee on National Parks, Forests and Public Lands, Washington, DC, April 29, 2021).

25. Law, "Wildfire in a Warming World"; Dominick A. DellaSala et al., "Pri-

mary Forests Are Undervalued in the Climate Emergency," *BioScience* 70, no. 6 (May 2020), http://doi.org/10.1093/biosci/biaa030.

26. Srijana Mitra Das, "Nature's Ecosystem Services Generate up to $140 Trillion per Annum—Significantly Greater than Global GDP: Onno van den Heuvel," *Economic Times* (India), May 18, 2023.

12 You Find Ecology in Everything Everywhere

Epigraph: James Lichatowich, *Salmon without Rivers: A History of the Pacific Salmon Crisis* (Washington, DC: Island Press, 1999), 227.

1. Andrea Wulf, *The Invention of Nature: Alexander von Humboldt's New World* (London: John Murray, 2015).

2. Eugene P. Odum, *Fundamentals of Ecology* (Philadelphia: W. B. Saunders, 1971); John McPhee, *Encounters with the Archdruid: Narratives about a Conservationist and Three of His Natural Enemies* (New York: Farrar, Straus, and Giroux, 1971); Aldo Leopold, "On a Monument to the Pigeon," in *Sand County Almanac* (New York: Oxford University Press, 1949), 101–4.

3. Lucy L. W. Owen et al., "Fractal Brain Networks Support Complex Thought," *Nature Communications* 12, 5728 (September 2021), https://doi.org/10.1038/s41467-021-25876-x.

4. Andrea Wulf, "The Forgotten Father of Environmentalism," *Atlantic,* December 2015, 421–34; Charles Darwin, *The Origin of Species* (1859; repr., New York: Signet, 2002), 559; this is the concluding paragraph of Darwin's most famous book.

5. William E. Odum, "Environmental Degradation and the Tyranny of Small Decisions," *BioScience* 32, no. 9 (October 1982): 728–29.

6. Brooke Penaluna, Lookout Fire update, presented to the H. J. Andrews scientific community, August 14, 2023.

7. Matt Betts, interview with the author, February 26, 2024.

8. Fred Swanson, email message to author, November 7, 2023; the poem is from Ariwara no Narihira (825–880), a Japanese poet and aristocrat.

9. Stanley A. Temple, "A Bird We Have Lost and a Doubt We Have Gained," *Wisconsin People and Ideas* 60, no. 3 (Summer 2014): 12–14. The last passenger pigeon died in 1914 alone in a cage at a Cincinnati Zoo.

10. Kratika Tandon, "The Birth of Environmental Law," *Q Magazine* 5, no. 2 (February 21, 2022): 42–47.

11. The White House, "Executive Order on Strengthening the Nation's Forests, Communities, and Local Economies," April 22, 2022, https://www.whitehouse.gov/briefing-room.

12. Odum, "Environmental Degradation."

13. Biodiversity Heritage Library, "Alexander von Humboldt and the Interconnectedness of Nature," October 2020, https://blog.biodiversitylibrary.org.

14. "Alexander von Humboldt," Stanford Encyclopedia of Philosophy, January 16, 2023, https://plato.stanford.edu.

15. Gregory N. Bratman et al., "Nature Experience Reduces Rumination and Subgenual Prefrontal Cortex Activation," *PNAS* 112, no. 28 (July 14, 2015): 8567–72; Ruth Ann Atchley et al., "Creativity in the Wild: Improving Creative Reasoning through Immersion in Natural Settings," *PLoS One* 7, no. 12 (December 12, 2012), http://doi.org/10.1371/journal.pone.0051474.

16. Kim Eckart, "What Counts As Nature? It All Depends," University of Washington News, November 15, 2017, https://www.washington.edu/news.

Postscript

Epigraph: Albert Einstein, *The Expanded Quotable Einstein,* ed. Alice Calaprice (Princeton: Princeton University Press, 2000), 247.

1. Alberto Cairo, "The Island of Knowledge and the Shoreline of Wonder: Using Data Visualization to Prompt Exploration," Neiman Lab, March 9, 2016, fig. 1.6, https://www.niemanlab.org. Alberto Cairo is Knight Chair in Visual Journalism in the School of Communication at the University of Miami.

2. Thomas S. Kuhn, *The Structure of Scientific Revolutions,* vol. 2, no. 2 (Chicago: University of Chicago Press, 1970), 202.

3. Fred Swanson, oral history, Oregon State University Archives, November 13, 2013, 00:03:00.

INDEX

global warming: and climate change, 193; and disturbance, 200; eight-hundred-thousand-year carbon dioxide record exploded, 192; and forests, 111; heat wave, 198–99; record-high in Pacific Northwest, 142; and temperatures, 114, 156; two degrees Celsius as upper limit of, 23; and water, 198. *See also* climate change
gold rush, 27, 29, 163
Grant, Gordon, 31, 36, 134, 151, 231

Halofsky, Joshua, 201
Hardin, Garrett, 63
Harmon, Mark, 81, 83–85, 172
hemlock: eastern, 157; hemlock woolly adelgid, 157; western, 17–19, 52–53, 67, 73–74, 95, 98–99, 102, 108–9, 111, 155, 157, 186, 230
herbicides, 57–58, 60
Herring, M. L.: childhood home, 64; coming of age, 48; ecological studies, 210–11; fieldwork, 128, 132, 212, 214, 216, 234; house fire, 206–7; house in Coast Range, 8, 29, 55–56, 60–61, 149, 175, 191, 212–13, 226; illustrations, x, 228, 233–34; missed owl war drama, 175; and Mount St. Helens, 43; neighbors' stories of giant old trees, 13, 227; neighbors' stories of old floods, 149–50; redcedar roof, 29, 120, 206–7, 213
Hill, Julia Butterfly, 180–81
H. J. Andrews Experimental Forest, 82, 93–95, 113–14, 150, 152, 167–73, 218–20, 232–34; Lookout Creek, 150, 166–67, 220, 232; mapping landforms beneath, 40; scientific study of old growth began at, 8; tree decomposition study, 84, 218; wildfire at, 219–20
huckleberry, 17, 54, 98, 102, 161
Hudson Bay Company, 27
humans: and air pollution, 203; carbon consumption, 192; and climate change, 157, 192–93, 196, 199, 206, 209; difficulty grasping ecology, 220–21; and disturbance, 45, 138, 193; ecosystem services of wild forests, 205–6; enjoyed climate stability for ten thousand years, 193; and environment, acceptance of, 223–24; and fire, 145, 191; human brain, 211–12, 215, 223; human ecosystems, 216; knowledge, 229, 231; and nature, 134–35, 148, 165, 167–68, 187, 211–12; and old growth, 63, 74–75; Pacific temperate rainforest inhabited for twelve thousand years, 160–61; senses, 100; smell, 99–100; society grapples with forests, 6–7; and temperate rainforest change, 11; tyranny of greed, 221; tyranny of small decisions, 216–18, 221, 225; tyranny of small imaginations, 221–22; understanding what it means, 13; vision, 97–99; walking, urban vs. forest, 223. *See also* European settlement; Native people
Humboldt, Alexander von, 210, 215, 222–23, 235

Ice Age, 37–38, 161
illustrations, x, 228, 233
imagination, 7, 39, 63, 97, 225–26; tyranny of small imaginations, 221–22
Indigenous people. *See* Native people
Ingalsbee, Tim, 146
insects: beetles, 49–50, 74, 83–84, 91, 93, 99, 128, 149, 153–55; and burned forest, 49, 55; and climate change, 155–56; and decomposition of trees, 83–84; endophytes ward off, 103–4; hemlock woolly adelgid, 157; meadows attract, 21; and old growth, 68; and streams, 116, 123–24, 128
International Biological Program (IBP), 167–73, 177, 187